The New York Times BOOK OF THE BRAIN

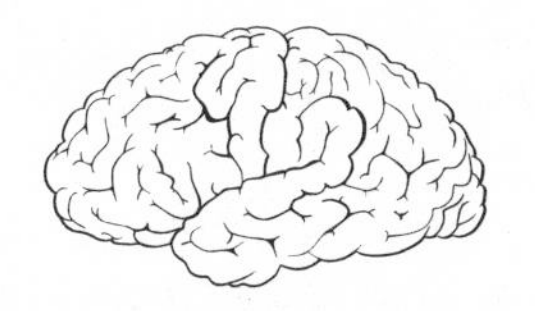

EDITED BY
NICHOLAS WADE

THE LYONS PRESS
GUILFORD, CONNECTICUT
An imprint of The Globe Pequot Press

Other books in the series:

The New York Times Book of Archeology
The New York Times Book of Birds
The New York Times Book of Fossils and Evolution
The New York Times Book of Genetics
The New York Times Book of Natural Disasters

Originally published under the title *The Science Times Book of the Brain*

The Lyons Press is an imprint of the Globe Pequot Press.

Printed in the United States of America

10 9 8 7 6 5 4 3 2 1

Designed by Joel Friedlander, Marin Bookworks

The Library of Congress Cataloging-in-Publication Data is available on file.

ISBN 1-58574-532-4

Contents

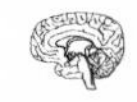

Introduction

ONE DAY SOMEONE will write a manual to the human brain. The manual will lay out how a new brain is wired up and what each of its components does. It will explain how the brain manages to generate within itself a remarkably faithful representation of the outside world, together with the consciousness of that representation. It will document the computational powers by which the brain accomplishes this feat in real time and also keeps a permanent running record of what it has done. It will explain how the instructions to build such a breathtaking machine are encoded into a mere six feet of computer tape, the DNA that is packed into the nucleus of each human cell.

Biologists who study the brain are many years away from being able to write such a manual. But they can at least now describe the nature of the enigma they are seeking to understand. And though the task remains daunting in many ways, the blossoming of new techniques has opened up an array of new approaches to the brain.

The articles in this book appeared in *The New York Times* over the last five years, a period during which neuroscience has become more interesting than ever. Written for the journalistic purpose of reporting fresh discoveries, they convey in this collection both the breadth of activity in brain research as well as a flavor of the tumult in the field. They may serve to prepare readers for the discoveries of the next few years as researchers close in on one of the last outstanding problems in biology.

The classical method of studying the brain is in essence to stick a recording electrode into a nerve cell and see what happens. This technique, still the basis of neurophysiology, has now been supplemented by powerful approaches at two quite different levels of complexity.

At the macro level, new imaging techniques for visualizing the living brain can display the behavior of whole populations of nerve cells, or neurons. The images, based on measures of the extra blood flowing to electrically active regions, have given neuroscientists vivid insights into the dynamics of the living brain as its different processing areas light up in response to a given task.

At the level below the neuron, the powerful techniques of modern genetics are helping to reconstruct the hereditary programs that underlie the cell's operation. The genes supply the information to wire up and operate the brain, and it is in the genes that ultimate explanations are to be found.

Study of the brain is a field awaiting grand unification. At present, molecular biologists explore events within the neuron, neurophysiologists figure out how neurons interact, and psychologists and psychiatrists study the behavior and pathology of the brain as a whole. When the brain is fully understood, the borders between these different disciplines will no doubt fade under the light of a more powerful synthesis.

Biologists' strategy is to try to understand the simpler functions of the brain first. Most progress has been made so far with the senses, the gateway to the brain, especially the sense of vision. The results so far tell us that the brain is very odd. It is a machine, but it works on principles quite alien to the engineer. The first thing the brain does with incoming sensory information is to deconstruct it. The eye immediately breaks down what it sees into elements of color, movement and shape. These elements are processed in separate neural tracts, then seemlessly recombined into the visual image of which the mind is aware.

Biologists have no understanding of the magic by which the final image is reconstituted. But the deconstruction suggests a lot about how the brain must operate. Its language is expressed in electrical signals and connections. All sensory input, whether sight, smell or touch, is converted into the same code, a train of electrical impulses in a given channel.

These impulses are processed subconsciously, with the mind being made aware only of the final result. Many other computations, such as setting the heartbeat and other housekeeping chores, never break into consciousness. The mind knows only what evolution has found advantageous to let it know.

The brain, after all, is an organ shaped by evolution for the specific purpose of favoring its owner's survival. It is much more than an electrical machine of pure thought. It is also a gland, seething with hormones that influence its own mood as well as the state of the body. Both the hormones and the thoughts are influenced by emotions, originating from hidden centers beneath the level of consciousness. In emergencies, thought is a luxury; the brain is hard-wired to direct our attention to things that may harm us. The eye perceives a moving shadow and your hair can be standing on end before you know it. Fear and other emotions tend to leap on us full grown, but presumably are in part the product of mental computation just as are our conscious thoughts.

Computation, by rules we have yet to fathom, is built into all the brain's operations. Consider the millions of bytes a computer would need to store all the elements of sight, sound and smell that are present in one conscious instant. The brain effortlessly processes the essential features, transfers them for intermediate storage to the hippocampus and files the relevant records for long-term storage somewhere back in the cortex. Here it accumulates the memories of a lifetime, most of them retrievable after a little thought.

The very oddest part of this extraordinary machine is that we should be aware of it. Biologists do not know to what degree, if at all, other animals are conscious. Our kind of consciousness, at any rate, seems to be a property of the human brain. Unfortunately, as the final chapter describes, we cannot yet explain it. Readers seeking an explanation of how their mind works will have to be patient a little while longer. Despite a profusion of learned books about the subject, consciousness is a mystery.

Those who write for newspapers expect their work to last but a single day. Together with my colleagues on the Science section of *The New York Times,* I am grateful to Lilly Golden and to Nick Lyons of the Lyons Press for granting our words an extension of life, one that the gathering momentum of brain research surely justifies.

—NICHOLAS WADE, Spring 1998

1

MAKING SENSE OF THE SENSES

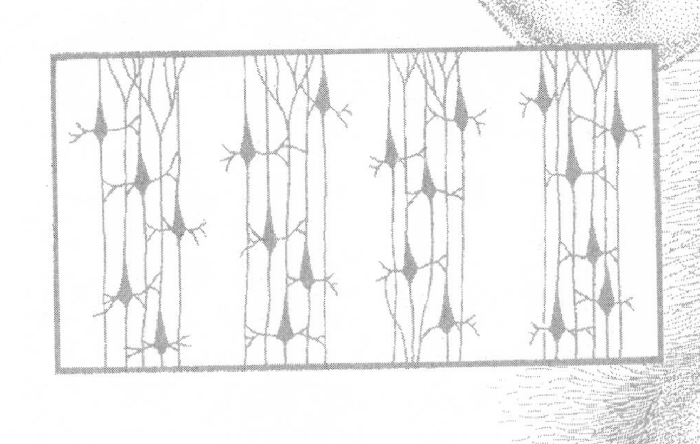

Sight, sound, smell, touch and taste—the senses are the doorways into the brain. Each converts its chosen category of information from the outside world into the electric signals that are the brain's common language.

The mystery of the brain begins at these very doorways. It is here that information is coded into the neural circuits. A ray of light becomes a signal in the optic nerve. A touch becomes a pattern of impulses in the spinal cord.

These sensory systems are the most accessible part of the brain and it is with them that many biologists have begun their studies. A reasonable understanding has been developed of how the brain sees, although many of the details remain obscure.

One principle that has emerged from the study of vision and the other senses is that information is highly processed before the brain becomes aware of it.

The eye does not work in any way like a camera, except in having a surface sensitive to light. Far from taking a snapshot of its field of view, the eye immediately breaks down what it sees into the separate categories of shape, movement and color. The information in these categories goes through further processing, according to rules built into the brain's visual circuits, and then, in ways that are not yet understood, is recombined. The result is the creation in the brain of a representation of the outside world.

Although we assume we are seeing the world directly, visual information goes through considerable computation before it reaches the conscious mind. The brain's task is to make sense of the world, and to this end it manipulates incoming visual information so as to enhance what is important and present a coherent scene. For instance, near the center of each eye's field of vision is a blind spot where nerve fibers are bundled and leave the retina as the optic nerve. Why isn't there a black spot in each eye's field of vision? Because the brain uses information from the other eye to compute what should be there and seamlessly fills the two spots in. Another clever piece of computation is that from the two-dimensional information gathered by each retina, the brain is able to construct a three-dimensional view of the world.

A second principle to have emerged from the study of vision is that sensory information is organized in the brain by means of maps. In the optic cortex, the part of the brain that deals with vision, there are many maps, sheets of nerve cells in which cells at one side correspond to the left part of the field of vision, and cells at the other side to the right. There is a similar topographic relationship between brain cells and sensory cells in the sense of touch.

The chapters that follow describe recent findings about the various senses and their operations.

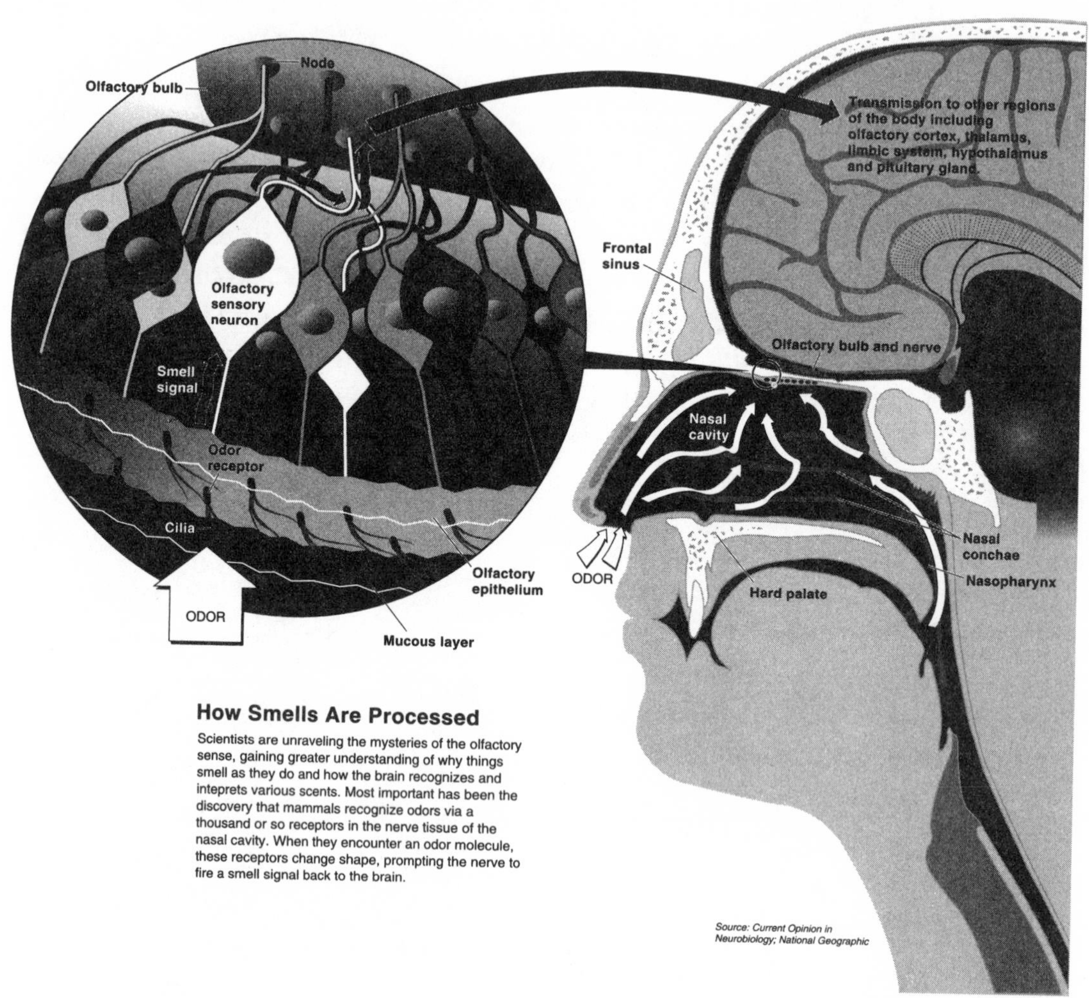

Illustration by John Papasian

Powerhouse of Senses, Smell, at Last Gets Its Due

ON THIS, the authorized day of passion and romance, when one tries hurriedly to spark a new flame or fan an old one back to life, when a wooer would prefer not to beat around the bush of love but rather dive right into it, the best Valentine's gifts may indeed be the staples: a dozen fragrant roses, a box of fine chocolates, a bottle of French perfume. After all, these items became shorthand for love by appealing shamelessly to the second-sexiest organ of the body—the nose.

Seeing may be believing and hearing may be music to the ears, but a good smell sails express non-stop to the deepest, warmest, happiest and most voluptuous sector of the self. And nothing can evoke a time, a place, an emotion past better than an aroma.

When Marcel Proust slipped into an eight-volume, orgiastic memory trance after eating a madeleine, it was not the little cake's taste that set him off so much as its smell. The tongue can distinguish only a handful of different inputs: sweet, sour, bitter, salty and possibly monosodium glutamate; detecting other flavors is a nose job.

Those who doubt the importance of smell to taste should try eating food while holding their nose, an exercise that will leave them utterly unable to distinguish a carrot from an apple, a Belgian truffle from a drugstore Brach's.

Like all mammals, humans can discern perhaps 10,000 or more odors. Smell is essential for most forms of human congress, particularly the most intimate forms. An infant finds its mother's nipple through the sense of smell, and a mother in turn can pick out her newborn based on its scent alone. A couple can survive all sorts of differences, notes Dr. Susan S. Schiffman, a professor of medical psychology at Duke University Medical Center, but if one dislikes the other's smell, the relationship is doomed. Many people prefer not to leave their private odor type to chance alone, which is why the perfume industry is a $6 billion-a-year enterprise. At the same time, the popularity of nouveau wellness techniques like aroma therapy is blooming.

As much as people recognize the profound role that smell plays in their affairs, the sense traditionally has not been taken seriously. "People snicker when they hear I work on smell," said Dr. Gary K. Beauchamp, director of the Monell Chemical Senses Center in Philadelphia. "They're also embarrassed by it." When patients complain of losing their eyesight or hearing, doctors pay attention; talk of a waning sense of smell, and doctors may joke that at least the subways will be more bearable.

Lately, however, the science of smell has been making enormous strides, propelled largely by the recent discovery that mammals recognize odors through the grace of a thousand or so distinct odor receptors located in the nerve tissue of the nasal cavity. The receptors are proteins that snake through the membranes of sensory nerves found in the so-called olfactory epithelium, a patch of tissue the diameter of two pencil erasers sitting right behind the bridge of the nose. Researchers believe that when an odor molecule floats through the air or is pushed upward from the throat during chewing or swallowing to meet the appropriate membrane receptor in the nose, the receptor changes shape and thus alters the property of the nerve cell. That change then prompts the nerve to fire a smell signal backward toward the brain.

The announcement of this giant family of odor receptors in 1991 was greeted with fanfare and surprise, for few had expected the smell system to be so elaborate and ungainly. The visual system, by contrast, can make out many thousands of hues through the use of just three color receptors, one tuned to red, another to blue, a third to green.

Since then, researchers have learned a great deal about how the odor receptors are distributed in the nose, how they are connected to the smelling centers of the brain, and how the receptors might operate singly or in teams to allow a person to discriminate between the scent of an orange and that of a lemon, a Crayola crayon and a bit of earwax, a roasting loin of pork and a kitchen curtain that has caught on fire.

In several papers in the journal *Cell,* Dr. Richard Axel, Dr. Robert Vassar and their co-workers at Columbia University College of Physicians and Surgeons in New York and Dr. Linda B. Buck and her colleagues at Harvard

Medical School in Boston independently described how odor information traveling from the nose to the brain becomes increasingly organized and refined. The detection of an odor begins with an arousal of smell receptors distributed more or less randomly within the nasal cavity. The information from those receptors is then focused onto elegantly structured "smell maps" within the all-important olfactory bulbs that serve as the relay center between nose and brain.

By studying the smell apparatus in rodents, the researchers have determined that each nerve cell in the olfactory epithelium holds only one type of receptor, but that there are many thousands of every type of nerve fiber—and thus of every type of receptor—scattered across the epithelium. But when those nerve cells send their connecting wires, or axons, back to the olfactory bulbs, all the axons projecting from the same class of fibers end up hitting a single node in one of the bulbs. In that way, the relative chaos of the odor-collection mechanism is reduced to military neatness of information within the bulbs.

The biologists also presented a model in which the smell system operates roughly like the immune system. According to this notion, the body confronts a new odor as it does a new microbe, that is, piecemeal. In attacking a pathogen, the immune system generates many different antibody proteins, each able to fix onto a small region of the enemy. So, too, might the smell system consider an odor from multiple angles, with one receptor recognizing one chemical signature of the odor molecule—say, a fatty acid chain of a particular length—while another receptor tunes in to a slightly different segment of the odor. The brain then synthesizes the disparate bits of information into a coherent label that experience has recorded as, for example, "smell of new plastic doll."

In other words, Dr. Buck said, there is probably no lemon-specific or rose-specific receptor in the nose. Instead, the smell of lemon is represented by the activation of a characteristic and presumably small series of receptors. And though similar odors probably stimulate overlapping groups of receptors, the patterns will vary enough to allow a person to distinguish among the lemon smell of a real lemon, the fake lemon scent of a gumdrop, and the version found in a lemon shampoo. With a thousand receptors to mix and match, the possible combinations for detecting smells are stupendously large, Dr. Axel said, amounting to 10 with 23 zeroes after it. However, he added, the human brain lacks sufficient neural connections to discriminate among that many odors, and so evolution has winnowed the network down to concentrate on classes of odors likely to have been relevant to ancestral human affairs, when the hominid nose was hammered into shape.

Thus, while the human odor-detecting apparatus is impressively versatile, able to pick up many new scents it has never confronted before, it cannot sniff out everything, including some common toxins that one might wish smelled as noxious as they are—carbon monoxide and natural gas, for example. Moreover, a human's olfactory might is 10 to 20 times less sensitive than that of a dog, and dimmer still than that of a rat.

Although much of the groundwork had been laid by electrophysiologists, anatomists and psychologists, the new molecular approach to studying smell has galvanized the field and given it a trendy cachet. The smell system offers researchers a unique window onto the brain. Unlike other nerve cells, which cannot regenerate, olfactory neurons die and are replaced throughout life, and scientists would love to learn how the smell cells accomplish what eludes the constituents of the spinal cord or neocortex.

The linkup between nose and brain likewise holds considerable interest. The olfactory bulbs extend some of their axons directly into the limbic system of the brain, the celebrated seat of emotions, sexuality and drive. Odor information thus goes from nose to bulbs to limbic system, a much more direct route than that traversed by visual and auditory input. Olfaction may be an ancient sense, perhaps the hoariest of them all; but its wiring to the brain is sweet and pithy.

Beyond illuminating the mechanism of smell, the new research may prompt greater appreciation and respect for the most neglected of the five senses. For other animals, smell reigns supreme.

If a female boar is ovulating, and she is given a sample of a male boar's urine to sniff, she will raise her haunches in anticipation of being mounted. Cats scent-mark their territory by releasing odor molecules from glands in their eyebrows, rumps and the pads of their paws.

Reporting in the journal *Science*, an international team of researchers demonstrated that when they altered the smell center of a male fruit fly's brain, the fly became bisexual, approaching male and female flies with indiscriminate ardor. The flies had very likely lost their ability to smell the signal that males emit to ward off other males.

Recent work from scientists in Belgium and at the Johns Hopkins University indicates that even sperm may rely on a kind of smelling method for wending their way toward an egg. Sperm cells turn out to bear on their surface the same odor receptors stippling the nerves of the olfactory epithelium.

Throughout the animal kingdom, the sense of smell is the link to life. "There are strains of mice that are blind and you would never know it from their behavior," Dr. Beauchamp said. "But there would never be a strain that couldn't smell. Such a mouse would not find food or a mate, and it would not persist."

Among humans, smell is perhaps less important for bodily survival. In fact, there are huge numbers of people who are anosmic, who lack a sense of smell, either from birth or following a head injury. Ben Cohen, a co-founder of Ben & Jerry's Homemade Inc., has said he is anosmic, which is why his ice cream is so rich in tactile and other sensory characteristics beyond flavor alone.

Yet while a lack of sense of smell may not be fatal, it can be depressing. "We had a patient in here who hit his head recently and lost his sense of smell," Dr. Schiffman said. "He's quite unhappy, he doesn't feel like eating, and he's lost 10 pounds."

As people age, they almost invariably suffer a decline in their sense of smell; developing a taste for spicy foods late in life is not unusual. Spicy food stimulates odor receptors, as all food does, by traveling from the back of the throat up to the olfactory epithelium. More to the point, it also has a "kick" that excites the trigeminal nerve, which controls the muscles of the face and jaw. Horseradish, alcohol, menthol—all have this tingly, so-called chemesthetic component that is independent of taste and smell, said Dr. Charles J. Wysocki of the Monell center.

As bad as anosmia is, sometimes being able to smell can be even worse. A persistent bad smell can be more debilitating than even the continuous cry of a car alarm. Dr. Schiffman and her colleagues are now studying a group of people in North Carolina who live downwind of hog farms. "They constantly have fecal odor in their kitchen and on their drapes," she said. "They can't open doors or windows."

"We've done profiles of their mood states," she added, "and we've found they're severely depressed, anxious and have less vigor."

Interestingly, though, what is foul and what is fair may depend largely on training. There are few smells that people universally rate as either good or bad.

Most cultures celebrate the sweetness of flowers, but people who work in the funeral business come to associate the odor with death and decay. Adults find the smell of feces to be highly objectionable, but young children do not, and perfumes are known to have their "fecal notes."

Skunk odor may be unbearable in high concentrations, but many consider a faint whiff of it to be almost pleasant. Skunk smell is not terribly different from the odor of musk and civet, essential ingredients in perfume that were originally isolated from glands in the abdomen of the male musk deer and near the anus of the African civet cat.

As it turns out, one of the only scents that appeals to people around the world and of all ages is the aroma of cola, prompting some scientists to suggest the cola preference may be innate. That could explain why, no matter where one wanders, no matter how remote the locale, a bottle of some variety of cola is never far from view.

—NATALIE ANGIER, February 1995

Brain Locates Source of a Sound with Temporal, Not Spatial, Clues

ROAMING THE WOODS on a lovely spring afternoon, a birdwatcher hears the tat-tat-tat of a woodpecker at work. His head turns toward the sound, which came from the left, probably near that big oak tree 50 yards away. Yes, there it is—a downy woodpecker.

Localizing interesting sounds is an ordinary human experience, but to scientists who study the human auditory system, the feat is mind-boggling. The woodpecker emits sound waves that travel through air, enter human ears and set off a chain reaction of impulses and computations that result in a person's being able to locate the bird's position, instantly and accurately. How this occurs remains cloaked in mystery.

Recently, a researcher in Florida turned up a new clue. Although his finding does not solve the larger puzzle, it does show that the auditory system has evolved coding strategies that work differently from other sensory systems. Moreover, the auditory cortex—the higher brain region where sounds are interpreted and understood—contains a population of cells that appear to be unlike any others in the nervous system.

The researcher, Dr. John C. Middlebrooks in the department of neuroscience at the University of Florida Brain Institute in Gainesville, published his work in *Science* magazine.

These special cells in the cortex can detect sounds occurring anywhere in the 360 degrees of space around the human body, Dr. Middlebrooks said. But they have developed a special code for differentiating the location of these sounds. For example, if a sound comes from, say, 10 feet to the left of a person's head, the cell will fire its signals in one pattern. "It might go dit, dit, ditty, dit," Dr. Middlebrooks said. But if the sound originates from a different point in space, the cell fires another pattern, he said, perhaps "ditty

ditty dit." All sounds from all locations can be coded in this way by these cells, he said.

"It's remarkable to show that a single neuron can encode sounds from every direction in space," said Dr. Eric Knudson, a professor of neurobiology at Stanford University medical school. It means that these auditory cells operate according to rules that are different from those governing other brain cells that help make sense of the outside world.

But the finding does not solve the question of how humans localize sounds, Dr. Knudson said. Although these cells produce a code, he said, it is not clear how or if the brain uses the signals in sensory processes. They could be passed on to other brain circuits to help understand sounds, Dr. Knudson said, or they could be an artifact of other processes.

The brain contains internal "maps" or representations of space that help people understand the world, Dr. Knudson said. Each sensory system—seeing, hearing, touching and so on—has specialized neurons that help create these maps, which are constantly being acted upon and re-created as humans react to the external world. (In dreams, the same cells are activated, but not by the outside world—the brain generates its own signals from within and tries to make sense of them.)

The neurons that perform these mapping functions have what are called receptive fields—a limited part of space to which they react. For example, the visual system contains cells that detect only horizontal lines. Others fire only when exposed to vertical lines. A neuron with a receptive field for horizontal lines will not fire signals to other neurons when it encounters vertical or diagonal lines or other basic shapes. It is activated only by horizontal lines—like the surface of a desk.

There are cells that specialize in color, motion, where the head is located in space and where the body is located in space, Dr. Knudson said. Some help determine what an object is and others help decide where it is.

Similar detailed maps exist for the somatosensory system—there are cells that tell people where their body parts are located. And there are maps for the motor system—cells that help calculate the trajectory of a baseball flying toward a person's hand.

All these maps are spatial representations of the outside world, said Dr. Knudson. These specialized cells, with their small receptive fields, then pass their mapped information up through various networks to the cortex—a thin layer of cells at the top of the brain where advanced processing takes place. Then, through intricate connections, the information is passed back down the same networks for the final act of cognition or perception.

This model has worked so well that researchers thought there should be spatial maps in the auditory system, Dr. Middlebrooks said. And, in fact, there are. The inner ear maps the frequency, timing and loudness of sounds. These signals are then sent to the midbrain where processing begins to determine sound location. Basically, the brain compares the intensity and timing of sounds received in each ear. If a sound arrives at one ear before the other, the time delay is used to help calculate where the sound is coming from. People with uneven hearing loss may have trouble, for this reason, locating sounds.

This information is then mapped onto another midbrain region, said Dr. Eric Brugge, a professor of neurophysiology at the University of Wisconsin in Madison. There, certain neurons represent points in acoustic space. A sound in one space causes one subset of neurons to fire, he said. When the sound moves, another subset of neurons will fire. Such topographic sound maps are superimposed on visual maps, Dr. Brugge said. But this all takes place in lower regions of the brain.

The auditory cortex is where sound identification occurs, Dr. Knudson said. "This is where we tell friend from foe." Is it a sound you want to run from? Or to move toward?

"We know the auditory cortex is involved in sound localization," Dr. Middlebrooks said. People who suffer a stroke in certain regions of the auditory cortex lose the ability to localize sounds on the opposite side of the body, he said. But sounds are processed over such a wide region of brain that this problem does not seriously interfere with a person's life, Dr. Middlebrooks said.

Given the many spatial maps in sensory systems, many researchers have been looking for similar maps in the auditory cortex, Dr. Middlebrooks said. There might be cells that map different aspects of hearing, like sound localization, sound motion and suppression of echoes.

But this is where his new research made a surprising finding. By placing electrodes in the brains of anesthetized cats, Dr. Middlebrooks and his colleagues found a population of cells that fire when sounds arrive from different locations. They did not map space in the conventional manner, he said. They are broadly tuned to sounds from all points in space and emit their signals in a temporal code rather than a spatial one.

Their role in auditory perception is not known, Dr. Middlebrooks acknowledged. Other cells may be "listening" to these codes but where they are located and what they might do with the code is unknown.

The brain may not need a spatial map for sound, said Dr. Antonio Damasio, a neuroscientist at the University of Iowa. Sensory systems create internal images of space at many brain levels, he said, and when sounds enter the brain, they may be referred to these spatial frames for interaction. Thus sound localization involves not only hearing the woodpecker but also turning the head and eyes toward the sound.

It makes sense that the auditory system would use a temporal map rather than a spatial one for decoding the world, Dr. Brugge said. Sounds occur in time, rather than space, and the brain has simply evolved a system for dealing with this aspect of nature, he said.

On the other hand, he said, temporal codes may be used by other sensory systems to tie spatial maps together. This idea, called synchrony, says that networks of neurons that carry out subspecialized tasks will all fire together for brief instants to give rise to complete perceptions. Once the perception is complete, the networks fall back into separate components where they can be reactivated in different combinations.

—SANDRA BLAKESLEE, May 1994

Seeing and Imagining: Clues to the Workings of the Mind's Eye

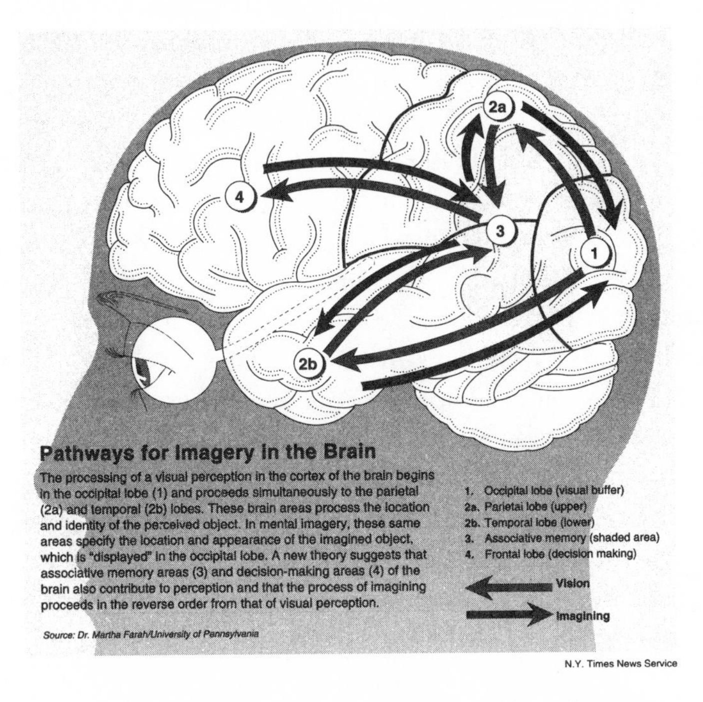

Pathways for Imagery in the Brain

The processing of a visual perception in the cortex of the brain begins in the occipital lobe (1) and proceeds simultaneously to the parietal (2a) and temporal (2b) lobes. These brain areas process the location and identity of the perceived object. In mental imagery, these same areas specify the location and appearance of the imagined object, which is "displayed" in the occipital lobe. A new theory suggests that associative memory areas (3) and decision-making areas (4) of the brain also contribute to perception and that the process of imagining proceeds in the reverse order from that of visual perception.

Source: Dr. Martha Farah/University of Pannsylvania

N.Y. Times News Service

CLOSE YOUR EYES and consider these questions: What shape are a German shepherd's ears? Which is darker green: a frozen pea or a Christmas tree? If you rotate the letter "N" 90 degrees to the right, is a new letter formed?

In seeking answers to such questions, scientists say, most people will conjure up an image in their mind's eye, mentally "look" at it, add details one at a time and describe what they see. They seem to have a definite picture in their heads.

But where in the brain are these images formed? And how are they generated? How do people "move things around" in their imaginations?

Using clues from brain-damaged patients and advanced brain imaging techniques, neuroscientists have now found that the brain uses virtually identical pathways for seeing objects and for imagining them—only it uses these pathways in reverse.

In the process of human vision, a stimulus in the outside world is passed from the retina to the primary visual cortex and then to higher centers until an object or event is recognized. In mental imaging, a stimulus originates in higher centers and is passed down to the primary visual cortex, where it is recognized.

The implications are beguiling. Scientists say that for the first time they are glimpsing the biological basis for abilities that make some people better at math or art or flying fighter aircraft. They can now explain why imagining oneself shooting baskets like Michael Jordan can indeed improve one's athletic performance. And, in a finding that raises troubling questions about the validity of eyewitness testimony, they can show that an imagined object is, to the observer's brain at least, every bit as real as one that is seen.

"People have always wondered if there are pictures in the brain," said Dr. Martha Farah, a psychology professor at the University of Pennsylvania. More recently, she said, the debate centered on a specific query: As a form of thought, is mental imagery rooted in the abstract symbols of language or in the biology of the visual system?

The biological arguments are winning converts every day, Dr. Farah said. The new findings are based on the notion that mental capacities like memory, perception, mental imagery, language and thought are rooted in complex underlying structures in the brain. Thus an image held in the mind's eye has physical rather than ethereal properties.

Mental imagery research has developed apace with research on the human visual system, said Dr. Stephen Kosslyn, a psychologist at Harvard University who is a pioneer in both fields. Each provides clues to the other, he said, helping to work out the details of a highly complex system.

Vision is not a single process but rather a linking together of subsystems that process specific aspects of vision. To understand how this works, Dr. Kosslyn said, consider looking at an apple on a picnic table 10 feet away. Light reflects off the apple, hits the retina and is sent through nerve fibers to an early visual way station that Dr. Kosslyn calls the visual buffer. Here the apple image is literally mapped onto the surface of brain tissue as it appears in space, with high resolution.

"You can think of the visual buffer as a screen," Dr. Kosslyn said. "A picture can be displayed on the screen from the camera, which is your eyes, or from a videotape recorder, which is your memory."

In this case, he said, the image of the apple is held on the screen as the visual buffer carries out a preliminary analysis of the scene. Edges, contour, color, depth of field and a variety of other features are examined separately, he said. But the brain does not yet know it is seeing an apple.

Next, distinct features of the apple are sent to two higher subsystems for further analysis, Dr. Kosslyn said. They are often referred to as the "what" system and the "where" system. The brain needs to match the primitive apple pattern with memories and knowledge about apples, he said, and so it seeks knowledge from visual memories that are held like videotapes in the brain.

The "what" system, in the temporal lobe, contains cells that are tuned for specific shapes and colors of objects, Dr. Kosslyn said. Some respond to red, round objects in an infinite variety of positions, ignoring local space. Thus the apple could be on a distant tree, on the picnic table or in front of your nose, he said, and it would still stimulate cells tuned for red round objects, which might be apples, beach balls or tomatoes.

The "where" system, in the parietal lobe, contains cells that are tuned to fire when objects are in different locations. If the apple is far away, one set of cells is activated, while another set fires if the apple is close up. Thus the brain has a way of knowing where objects are in space so the body can navigate accordingly.

When cells in the "what" and "where" systems are stimulated, they may combine their signals in yet a higher subsystem where associative memories are stored, Dr. Kosslyn said. This system is like a card file where visual memories, as if held on videotapes, can be looked up and activated, he said.

If the signals from the "what" and "where" systems find a good match in associative memory, Dr. Kosslyn said, you know the object is an apple. You also know what it tastes and smells like, that is has seeds, that it can be made into your favorite pie and everything else stored in your brain about apples.

But sometimes, he said, recognition does not occur at the level of associative memory. Because it is far away, the red object on the picnic table could be a tomato or an apple. You are not sure of its identity, he said. So another level of analysis kicks in.

This highest level, in the frontal lobes, is where decisions are made, Dr. Kosslyn said. To use the same analogy, it is like a catalogue for the videotapes in the brain. You look up features about the image to help you identify it, he said. A tomato has a pointed leaf, while an apple has a slender stem. When the apple stem is found at this higher level, he said, the brain decides that it has an apple in its visual field.

Signals are then fired back down through the system to the visual buffer and the apple is recognized. Significantly, Dr. Kosslyn said, every visual area that sends information upstream through nerve fibers also receives information back from that area. Information flows richly in both directions at all times.

Mental imagery is the result of this duality. Instead of a visual stimulus, a mental stimulus activates the system, Dr. Kosslyn said. The stimulus can be anything, including a memory, odor, face, reverie, song or question.

For example, Dr. Kosslyn said, "I ask you to imagine a cat." Images are based on previously encoded representations of shape, he said, "so you look up the videotape in associative memory for cat."

When that subsystem is activated, he said, a general image of a cat is mapped out on the screen, or visual buffer, in the primary visual cortex. It is a stripped-down version of a cat and everyone's version is different.

"Now I ask you another question," Dr. Kosslyn said. "Does the cat have curved claws?" To find out, the mind's eye shifts attention and goes back to higher subsystems where detailed features are stored.

"You activate the curved claws tape," he said, "and then zoom back down to the front paws of the cat and you add them to the cat. Thus each image is built up, a part at a time."

The more complex the image, the more time it takes to conjure it in the visual buffer, Dr. Kosslyn said. On the basis of brain scans with the technique known as positron emission tomography, he estimates that it requires 75 to 100 thousandths of a second to add each new part.

The visual system maps imagined objects and scenes precisely, mimicking the real world, Dr. Kosslyn said. "You scan it and study it as if it were there."

This can be demonstrated when people are asked to imagine objects at different sizes. "Imagine a tiny honeybee," Dr. Kosslyn said. "What color is its head?" To do this, people have to take time to zoom in on the bee's head before they can answer, he said.

Conversely, objects can be imagined so that they overflow the visual field. "Imagine walking toward a car," Dr. Kosslyn said. "It looms larger as you get closer to it. There comes a point where you cannot see the car at once. It seems to overflow the screen in your mind's eye."

People with brain damage often demonstrate that the visual system is doing double duty, Dr. Farah said. For example, stroke patients who lose the ability to see colors also cannot imagine colors.

An epilepsy patient experienced a striking change in her ability to imagine objects after her right occipital lobe was removed to reduce seizures, Dr. Farah said. Before surgery, the woman estimated she would stand, in her mind's eye, about 14 feet from a horse before it overflowed her visual field. After surgery, she estimated the overflow at 34 feet. Her field of mental imagery was reduced by half, Dr. Farah said.

Another patient suffered damage to his "what" system while his "where" system was intact. "If you ask him to imagine what color is the inside of a watermelon, he does not know," Dr. Farah said. "If you press him, he might guess blue. But if you ask him, is New Jersey closer to Oklahoma or North Carolina, he answers correctly instantly."

Imaging studies of healthy brains produce similar findings, Dr. Farah said. When a person is asked to look at and then to imagine an object, she said, the same brain areas are activated. When people add details to images, they use the same circuits used in vision. Interestingly, people who say they are vivid imagers show stronger activation of the relevant areas in the brain, she said.

People use imagery in their everyday lives to call up information in memory, to reason and to learn new skills, the scientists said.

It can lead to creativity. Albert Einstein apparently got his first insight into relativity when he imagined chasing after and matching the speed of a beam of light.

It can improve athletic skills. "When you see a gifted athlete move in a particular way, you note how he or she moves," Dr. Kosslyn said, "and you can use that information to program your own muscles." Basically, the brain uses the same representations in the "where" system to help direct actual movements and imagined movements, he said. Thus, refining these representations in imagery will transfer to actual movements, provided the movements are physically practiced.

Humans exhibit vast individual differences in various components of mental imaging, which may help explain certain talents and predilections, Dr. Kosslyn said. Fighter pilots, for example, can imagine the rotation of complex objects in a flash, but most people need time to imagine such tasks.

In a study in progress, Dr. Kosslyn and colleagues are examining the brains of mathematicians and artists with a new imaging machine that reveals individual differences in the way brains are biologically wired up. They are looking to see if people who are good at geometry have different circuitry from that of people who are good at algebra.

In a philosophical conundrum arising from the new research, it seems that people can confuse what is real and what is imagined, raising questions about witnesses' testimony and memory itself.

"In visual perception," Dr. Kosslyn said, "you prime yourself to see an object when you only have part of the picture. If you expect to see an apple, its various fragments can drive the system into producing the image of an apple in your visual buffer." In other words, he said, you prime yourself so much that you actually play the apple tape from your memory banks.

Thus, people can be fooled by their mind's eye, Dr. Kosslyn said. Imagine seeing a man standing before a frightened store clerk and you assume a robbery is under way. It is dark and his hand is in the shadows. Because you expect to see a gun, your thresholds are lowered and you may actually run the tape for a gun, even though it is not there. As far as your brain is concerned, it saw a gun, Dr. Kosslyn said. Yet it may not have been real.

Luckily, he said, inputs from the eye tend to be much stronger than inputs from imagination. But on a dark night, under certain circumstances, it is easy to be fooled by one's own brain.

It is amazing that imagination and reality are not confused more often, said Dr. Marcia Johnson, a Princeton psychologist who in her laboratory can make people swear that they saw or heard things that never happened. In general, she said, images are fuzzier and less coherent than real memories, and humans are able to differentiate them by how plausible they seem.

—SANDRA BLAKESLEE, August 1993

Brain Yields Clues to Its Visual Maps

A Theory of How the World Exists in the Brain

Every sight that enters the eye triggers the firing of neurons in the brain's cortex. In monkeys, researchers find distinct regions containing cells that fire only in response to particular characteristics: shapes, even specific shapes like lips; an object's angle; motion to the right or left, up or down; and movement within textured fields, like grass or leaves. Regions that respond to particular sights have been called "maps."

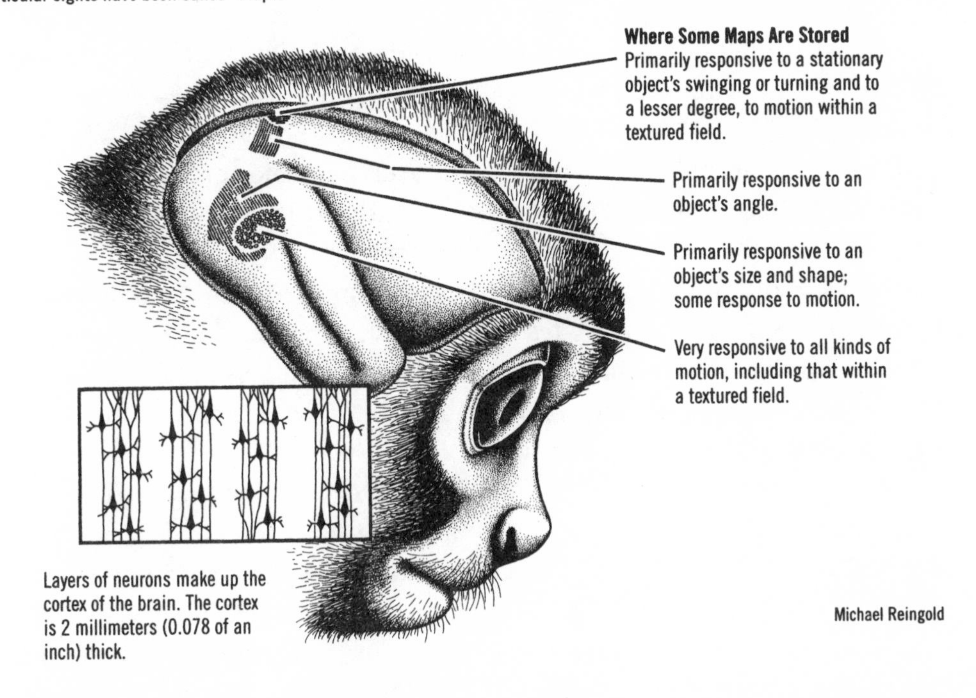

Michael Reingold

AMID THE VAST VARIETY of human faces, many very similar in appearance, how can an individual instantly recognize those few that are familiar, even ones not seen for many years? The answer may lie in a newly discovered alignment of brain cells that is believed to form the basis of visual memory.

The cells, located in the brain's last and most sophisticated stop for processing visual information, were shown to be stacked in arrays of columns. By studying the firings of individual nerve cells in this region, the researchers found that each cell in a given column responded to very similar but slightly different images, and each neighboring column of cells responded to another closely related sequence of images.

Depending upon exactly which columns are "excited" by a particular face and exactly which cells in each column are turned on by the various elements of that figure, the brain may be able to distinguish person from dog, woman from man, friend from stranger.

The finding, by Japanese neurophysiologists who described it in the journal *Nature,* is but one of many recent discoveries in the neuromechanics of vision that are beginning to explain how the human mind can instantly recognize complex images like faces even at odd angles or when only part of the face is visible.

Other studies of basic visual mechanisms are delving into how the eye and the brain magnify incoming light so that animals can see clearly both day and night.

The foundation for the Japanese discovery was laid through studies of soldiers who received brain injuries in the Russian-Japanese War of 1904. By correlating the soldiers' injuries with their exact visual losses, neurophysiologists discovered where in the brain the images from the retina were represented. Since then, many researchers have been developing a detailed brain map for visual information. Among these researchers are Dr. John Allman of the California Institute of Technology in Pasadena and Dr. Jon Kaas of Vanderbilt University in Nashville.

By placing microelectrodes on individual cells in the brain, these scientists determined that different brain areas are sensitive to different kinds of visual input: color, various aspects of motion like direction and speed, size, depth and shape.

In an interview, Dr. Allman said he and Dr. Kaas had so far "found an enormous number of maps—25 or more—outside the primary visual cortex." The primary visual cortex is a region on the pleated surface of the brain, at the back of the head, where visual information is processed.

He said there was reason to believe from animal studies that when one visual map was damaged, another could take over its function, suggesting that the brain may be able to compensate to some extent for certain kinds of vision-impairing damage.

The number of visual maps an animal has is related to how heavily it depends upon vision, Dr. Allman said. "But the great mystery," he said, "remains how all these maps are put together to tell the animal what it is seeing: how the brain learns to see new things and how it pulls meaningful information out of confusing images."

Earlier research had shown that columns of cells in the primary visual cortex are highly selective for a particular visual stimulus. Each cell sends messages to other cells only when this stimulus, such as a certain kind of edge, is perceived. Recent studies have focused on how the brain links a variety of different stimuli to form an image of a whole and how one image triggers memory of another.

Monkey studies described in the journal *Nature* by Kuniyoshi Sakai and Yasushi Miyashita at the University of Tokyo revealed that individual brain cells selective for different complex stimuli are able to link stimuli that are presented in quick succession, just as frames of film are joined to make a movie.

Dr. Michael Stryker, a neurobiologist at the University of California at San Francisco who wrote a commentary on the research, believes this may be the foundation of short-term visual memory that allows people and animals to make sweeping visual generalizations from a particular image. It is how an animal can conclude that an object is what it is even when only a part of it is perceived or when it is seen from a different angle.

Thus, when a monkey sees a tiger, one group of brain cells may respond to the stripes, another to the two blobs that form the head and body, another to the stick-like components that form the legs. Once the brain has associated these characteristics, it may have to see only one of them to "recall" the others.

The existence of cells that respond only to certain very complex stimuli, like a hand or a face, was discovered by accident in a "failed" experiment by Dr. Charles Gross, now at Princeton University. Dr. Gross was trying to record electrical signals fired from individual cells in the cortex of an anesthetized rhesus monkey, but the cells were stubbornly unresponsive to simple stimuli like lines, bars or lights. Ready to abandon the experiment, Dr. Gross waved his hand in front of the setup in disgust and much to his astonishment the cells suddenly fired.

Vision researchers have dubbed these "grandmother" cells, presumably because one cell could recognize an image as complex as a grandmother's face. Similar super-recognition cells for special, important images have recently been demonstrated in animals like sheep, which have visual brain cells that respond, for example, only to another sheep or to a particular predator like a dog or man.

The newest Japanese studies, by Dr. Ichiro Fujita and colleagues at the Riken Institute in Wako, Saitama, examined the organization of cells and communication between them in the brain's "final station" for recognizing objects that the eyes see. This region, called the anterior inferotemporal cortex, lies on the side of the head in front of the ear.

According to Dr. Stryker, who also wrote a commentary on these studies, the scientists first tested the response of monkeys' brain cells to "a kindergarten full of toys—stuffed animals, artificial fruits and vegetables, and the like." Then step by step the researchers refined the image to which each cell on the surface of the cortex responded until a "basic" stimulus was found, like a black circle with a white center or a striped bar.

Using fine microprobes and steady hands, the researchers one by one tested the response of the cells beneath the first layer, then beneath the second layer, and so on, and showed that each cell in a column responded best to a stimulus that differed slightly from the one that excited the cell at the top of the column.

"The experiments suggest that a moderate-sized collection of almost iconic figures may be the alphabet in which our visual memories are written," Dr. Stryker wrote.

Judging from the size of the brain area involved and the responses and spacing of the cellular columns in it, he surmised that the alphabet may have at most only a thousand different "letters" from which complex visual images are constructed.

Between an image's arrival at the retina and its being recognized in the brain's visual cortex lie many steps. The very first, the eye's ability to perceive a barrage of photons from an image of interest, turns out to be so important to survival that eyes of one sort or another have evolved independently more than a dozen times, according to a report in *The Annual Review of Neuroscience* by Dr. Michael F. Land of the University of Sussex in Brighton, England, and Dr. Russell D. Fernald of Stanford University.

The two scientists wrote, "There is a relatively small number of ways to produce an eye that gives a usable image, and most have been 'discovered' more than once, giving rise to similar structures in unrelated animals." Thus, squid and fish, which are not evolutionarily related, have very similar eyes, but humans and fish, which have a common ancestor, are optically very different.

All manner of eyes, however, seem to rely on one class of protein called opsins to capture the basic element of light and translate it into an electrical signal that can be understood by brains, however primitive they may be.

Studies at Stanford and elsewhere have shown that opsin picks up incoming photons that in turn trigger electronic shifts within the attached chromophore. A resulting cascade of adjacent transducer proteins amplifies the energy in a photon of light by almost a million times, enabling a rod cell in the retina to detect it, Dr. Fernald said.

Another kind of amplification problem afflicts animals that must see well at night, when illumination of the earth is diminished by a factor of about one million. Recent studies have shown that sensitivity to incoming light in many animals is regulated by a biological clock that resides in the brain and sometimes in the eyes themselves.

One such animal is the horseshoe crab, a living fossil that, although it evolved 350 million years ago, has a surprisingly sophisticated visual system made up of 1,000 huge photoreceptor units. Dr. Robert B. Barlow, Jr., a neuroscientist at Syracuse University, has been studying these crabs for more than a quarter century, first proving that male crabs locate their mates on dark beaches at night by vision alone and then showing that a brain-driven clock turns up the animal's visual sensitivity at dusk by a millionfold, enabling it to see almost as well at night as in the day.

Dr. Barlow recently showed that the eyes of Japanese quail are also night-adapted by a biological clock, but in this case, as in rats and probably all vertebrates, the clock is in the eye itself. If the crab's optic nerve is cut, the brain cannot signal the eye and no change in sensitivity occurs at nightfall. But covering the crab's eyes does not block the day-to-night switch.

In vertebrates, the opposite is true: Cutting the optic nerve does not interfere with the dark-induced increase in visual sensitivity but covering the eyes does, indicating that the eye itself, not the brain, controls the visual circadian rhythm.

Dr. Barlow is now pursuing the $64,000 question in vision: "Just what is the neural code by which information is transmitted from the eye to the brain? It is as if there were 1,000 telephone calls coming into the brain at once. How does the animal's brain know what it is seeing?"

With photoreceptor units large enough to see with the naked eye, the horseshoe crab is an ideal study subject. Dr. Barlow said: "We can record signals from each of several photoreceptors while the animal is moving around underwater and compare what we record with a computer prediction based on simulated images from an underwater camera that sees what the crab sees. If what we record from the animal matches the computer prediction, then we know we're beginning to crack the code."

—JANE E. BRODY, March 1993

A Separate Pathway Slowly Carries a Caress to the Brain's Attention

RESEARCHERS HAVE DISCOVERED the scientific basis for enjoying a caress.

People, it seems, have a special pathway of nerves that send pleasure signals to the brain when the skin is gently stroked. The pathway is present at birth and may help human infants distinguish comfort from discomfort.

The new pathway was described last week by Dr. Hakan Olausson, a neurophysiologist at Goteborg University in Sweden, at the annual meeting of the Society for Neuroscience in Miami Beach, Florida.

Finding a sensory system for pleasurable caresses was a complete surprise, he said in an interview. The discovery could help explain symptoms experienced by people with peripheral nerve damage that affects the pathway, he said.

It has long been known that humans have separate nerve networks for detecting pain, temperature and touch, Dr. Olausson said. Such nerves are arrayed throughout the muscles and skin and transmit a message to the brain when stimulated.

Each nerve has more than 1,000 fibers that help pick up signals from receptors in the skin and muscles, Dr. Olausson said. The new discovery was made with a technique that involves inserting a very thin needle into a single nerve fiber and recording its activity. "We wanted to find out how the direction of a movement over skin was encoded by nerves," Dr. Olausson said. When the researchers inserted such needles into a large nerve just above the inside of the elbow of human volunteers, "we found a surprise," he said. Some fibers were much slower to respond to touch than expected.

It had been believed that all human nerves involved in touch send their signals to the brain at a fast speed, 200 feet per second, Dr. Olausson said. But these fibers sent their messages at three feet per second, he said. The researchers found that one third of the fibers in the touch-sensitive nerves responded at low velocity to gentle stroking.

Most animals have fast and slow pathways for responding to touch, Dr. Olausson said. But scientists had long believed that humans, in the course of evolution, lost the slow pathway for touch and retained a slow pathway for pain, he said.

Thus "if you spill hot water on your foot, you first feel a touch sensation followed by a pain sensation," Dr. Olausson said. The touch signal reaches the brain in a few hundredths of a second while the pain signal reaches the brain a second or two later.

The reason for the different signaling rates lies in a fatty sheath called myelin that encompasses some nerves and not others, Dr. Olausson said. Fast-acting nerves have myelin and slow nerves do not, he said. It was felt that fast-acting nerves are a later evolutionary step in nervous-system development.

But it now appears that humans have a slow pathway for touch, Dr. Olausson said.

To find out its purpose, the researchers blocked only the fast-acting fibers involved in nerves responding to touch. "When that happens, no sensation of touch remains," Dr. Olausson said. "You don't feel anything," even though the slow-acting fibers in that nerve are transmitting signals to the brain.

Conversely, the researchers found experiments in the scientific literature in which slow-moving pain pathways were selectively blocked. The structure of pain fibers and slow-touch fibers are similar, Dr. Olausson said. If pain is blocked, he said, slow-touch should also be blocked.

But when the slow-touch path is blocked, he said, and the subject is gently stroked, he or she feels an uncomfortable burning and itching sensation. It appears that the blocking of slow-touch pathways leads to a distorted touch sensation, an unpleasant feeling, Dr. Olausson said.

This led him to the idea that slow fibers must work with fast fibers to produce pleasurable sensations such as caresses. The slow-touch system may somehow also promote positive emotions, he said.

Dr. Olausson said that these conclusions are tentative and that experiments are under way to explore both the fast- and slow-acting fibers involved in touch. But a huge mystery remains. No one knows how these signals are interpreted by the brain as feelings of pleasure or pain.

—Sandra Blakeslee, November 1994

How Do You Stop Agonizing Pain in an Arm That No Longer Exists? A Scientist Does It with Mirrors

THE THIRD TIME he lost his left arm, Derek Steen yelped with joy. He could not believe his good fortune.

The problem began 10 years ago when Mr. Steen crashed his motorcycle, tearing all the nerves that attached his left arm to his spine. The arm was hopelessly paralyzed, bound in a sling. He lost his arm a second time, so to speak, one year later when, deemed useless and getting in the way, the limb was amputated.

But Mr. Steen, now a 28-year-old part-time worker in San Diego, continued to suffer. His phantom arm felt paralyzed, pressed against his body, and it ached horribly for 10 years.

Then, late last year Mr. Steen was cured. After a simple, three-week treatment involving mirrors, his paralyzed phantom arm suddenly vanished, as did the gnawing pain in his phantom elbow. In its place, Mr. Smith says, he has a much reduced phantom limb composed of his lower left palm and all five fingers, which he can now "wiggle" freely from the stump below his shoulder.

He is ecstatic, said Dr. Vilayanus S. Ramachandran, a professor of neuroscience at the University of California at San Diego, who devised the treatment. Mr. Steen's "body image," as he puts it, has been profoundly altered, he said, much for the better.

Dr. Ramachandran described the finding—a rare example of successful treatment for phantom limb pain—at the second annual meeting of the Cognitive Neuroscience Society being held in San Francisco. The three-day meeting drew nearly 700 scientists who are engaged in basic research on the nervous system and how it leads to cognitive acts like learning, remembering, perceiving and consciousness.

Phantom limbs occur when the brain modifies its sensory maps after an amputation, he said. The brain region mapping an arm no longer gets input from the arm but it continues to be constantly stimulated by inputs from adjacent body parts. These stimuli fool the brain into thinking the arm itself is still there.

But sometimes, as in Mr. Steen's case, a phantom limb can be paralyzed. "The arm is in a fixed position, as if frozen in cement," Dr. Ramachandran said. "The patient can't generate the slightest flicker of movement, even though he can feel all the parts. I started thinking why."

The answer lies in the feedback loops in the brain that integrate vision, senses, body movements, body image and motor commands, Dr. Ramachandran said. In the first few weeks after the accident, Mr. Steen would try to move his injured arm, to no avail. His brain sent signals to his arm, commanding motion, but though his eyes confirmed the arm was there, it did not move.

In time, Dr. Ramachandran said, Mr. Steen developed "learned paralysis."

"His brain constantly got information that his arm was not moving," even though it was still there, Dr. Ramachandran said. After the amputation, he still felt it was there.

"Now if it's true paralysis can be learned, can you unlearn it?" Dr. Ramachandran said. "And how do you do that if you don't really have an arm?"

Simple. With mirrors. Dr. Ramachandran constructed a simple box without a lid and front and placed a vertical mirror in the middle. By placing his right arm into the box, Mr. Steen could see a mirror image of his missing left arm.

"I asked him to make symmetric movements with both hands, as if he were conducting an orchestra," Dr. Ramachandran said. "He started jumping up and down and said, 'Oh, my God, my wrist is moving, my elbow is moving!' I asked him to close his eyes. He groaned, 'Oh no, it's frozen again,'" the scientist said. "The box cost only $5, so I told him to take it home and play around with it."

Three weeks later, Dr. Ramachandran said, "he phones me, sounding agitated and excited." The conversation went like this:

"Doctor, it's gone!"

"What's gone?"

"My phantom arm is gone."

"What?"

"All I have is fingers and a lower palm dangling from my shoulder."

"Does this bother you?"

"No," Mr. Steen replied. "The pain in my elbow is gone. I can move my fingers. But your box doesn't work anymore."

The telescoping of fingers is common in phantom limbs, Dr. Ramachandran said. "But nevertheless I couldn't but help think that I had permanently altered someone's body image."

The reason the arm disappeared and the body image changed probably has to do with tremendous sensory conflict, Dr. Ramachandran said. "His vision was telling him that his arm had come back and was obeying his commands. But he was not getting feedback from the muscles in his arm. Faced with this type of conflict over a protracted period, the brain may simply gate the signals. It says: 'This doesn't make sense. I won't have anything to do with it.'

"In the process, the arm disappears and the elbow pain goes away," he said. "The reason the fingers survive and dangle from the shoulder is that they are overrepresented in the cortex, much more so than the rest of the arm. So there may be a kind of tip-of-the-iceberg phenomenon going on here."

Dr. Ramachandran is now treating other kinds of phantom limb pain, including a clenching spasm of phantom hands. "People feel as if their fingernails are digging into their hands and say it is excruciatingly painful," he said. The mirror box has helped one such patient.

The new finding may also have relevance to stroke rehabilitation, Dr. Ramachandran said. In the early stages, some paralysis is due to swelling in the brain and learned paralysis could result. While destroyed tissue could not be revived, he said, other circuits might be re-established with the use of the mirrors.

These ideas remain highly speculative, Dr. Ramachandran said. Moreover, pain is notoriously susceptible to placebo effects, so it may be difficult to prove that these brain-based therapies are effective.

Nevertheless, striking results are being seen with some stroke patients who lose the ability to talk. At the University of Iowa, Dr. Antonio Damasio and his colleagues are teaching American Sign Language to patients whose primary language areas have been partly damaged but not destroyed. By learning a sign for a concept, he said, they are often able to reconnect with the word for that concept—and in this way can learn how to speak once again.

Such plasticity or adaptability in the adult brain may be more common than most people realize, said Dr. Michael Gazzaniga, a neuroscientist who is an expert on split brain patients, then at the University of California at Davis.

An effective treatment for severe epilepsy is to cut the bundle of fibers that connect the left and right brain hemispheres. Each half brain is conscious but does not know what the other half sees or does.

In such patients, the left brain "wakes up and starts talking," Dr. Gazzaniga said. The right brain usually shows no sign of language comprehension and it certainly has no ability to talk—until a patient named Joe stunned researchers by learning how to talk with his right brain. Joe was like other split brain patients for 13 years, Dr. Gazzaniga said. Then his right brain started recognizing and saying words independently of the left brain. Now, 15 years after surgery, Joe can name 60 percent of the stimuli presented to his right brain, Dr. Gazzaniga said.

"It could be that new connections have been made, or that others have been unrepressed," Dr. Gazzaniga said. "At this point we have no idea how or why it happens." But answers, should they be found, might be used for helping people overcome all sorts of brain injuries and neurodegenerative diseases.

—SANDRA BLAKESLEE, March 1995

2

THE MACHINERY OF MOOD AND EMOTIONS

Unlike the senses, which present a defined subject of study, the brain's moods and emotions are much harder to scrutinize.

Emotions are often expressed both in the brain and the body—a consciousness of rage and the trembling of the limbs—but the brain centers that coordinate these reactions operate beneath the level of consciousness, making them harder to study.

Traditional methods have identified regions of the brain that specialize in handling emotions, and new machines for imaging the brain at work have helped to further delineate these regions and their interactions. Recent research has centered on the amygdala, a small region that lies deep within each cerebral hemisphere and serves as the principal coordination center for many types of emotion.

The amygdala has two-way connections with the higher centers of the brain in the cerebral cortex. It is also connected with the hypothalamus, a central region that controls many aspects of the body's physiological state, such as heart rate and hormonal response. The amygdala's connections thus make it a good candidate for the role of the hidden puppet master that coordinates the emotional responses of brain and body.

A complete description of mood and emotion is far beyond present reach. Still, substantial progress has been made on many fronts, and the following chapters show how far scientists have gone in understanding the chemistry of mood and the connections in the brain whereby emotion is generated and expressed.

Tracing the Brain's Pathways for Linking Emotion and Reason

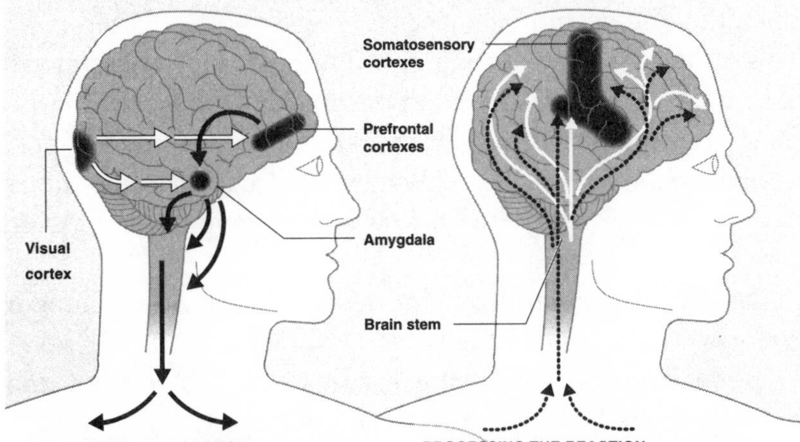

N.Y. Times News Service
Illustration by Baden Copeland

IMAGINE WALKING ALONG a jungle path in the twilight and hearing a lion roar. Your skin turns clammy, a knot forms in your stomach and you can taste the fear rising in your throat.

Now imagine walking along a zoo path at the same time of the evening and hearing the same sound. This time you do not feel afraid.

The reason, scientists say, has to do with how emotions and feelings are processed in the brain. External sensations (the roar) and memories (lions are locked up in the zoo) interact along complex circuits to generate our emotional reactions—in this case, to not be afraid.

Those neural circuits—actual networks of cells that crisscross the brain and send projections throughout the body—are now being described in unprecedented detail by a handful of neuroscientists who say the biological nature of emotions and feelings can at last be described.

Until recently, brain researchers focused most of their attention on the biological basis of cognitive processes such as perception and memory, said Dr. John Allman, a professor of neurobiology at the California Institute of Technology. They tended to ignore emotion, he said, in the belief that emotions and rational thought are separate activities and that emotions are just too difficult to understand biologically.

This attitude is now changing, Dr. Allman said, as researchers have come to realize that emotional brain circuits are just as tangible as circuits for seeing, hearing and touching. In this view, emotions and feelings are not, as poets and philosophers say, ephemeral reflections of the human soul. Rather, emotions are largely the brain's interpretation of our visceral reaction to the world at large.

Pioneering experiments on emotions have turned up some interesting concepts:

- Emotional memories involving fear are permanently ingrained into the brain; they can be suppressed but never erased.
- The body, as represented in the brain, is the frame of reference for what humans experience as mind. Our thoughts and actions—our sense of subjectivity—uses the body as a yardstick.
- Emotions are an integral part of the ability to reason. While too much emotion can impair reasoning, a lack of emotion can be equally harmful.
- Gut feelings and intuition are indispensable tools for rational decision-making; without them humans would have great difficulty thinking about the future.

Much of the new information about the neural circuits underlying emotion stems from experiments on animals. Dr. Joseph LeDoux, a professor of neurobiology at New York University and a pioneer in such research, said that a basic emotion like fear and the circuits that support its expression were highly conserved through evolution. Understanding fear mechanisms in animals, he said, sheds light on human fears and may help researchers study other emotions. The work is important because many psychiatric disorders, including anxiety, phobias, post-traumatic stress syndrome and panic attacks involve malfunctions in the brain's ability to control fear, he said.

Much of the research is centered on the amygdala, a tiny structure deep in the brain that is crucial for the formation of memories about significant emotional experiences. Damage a rat's amygdala and it "forgets" to be afraid.

To trace the cell networks involved in fear, Dr. LeDoux and his colleagues first conditioned rats by pairing a loud noise with a mild electric shock to their feet. The rats soon showed fear when they heard the noise without the shock. The researchers presume fear conditioning occurs because the shock modifies the way in which neurons in several brain regions interpret the sound of the stimulus.

In time, however, the rats gradually lost their fear of the sound. Some part of the rat's brain outside the amygdala seems to control the fear response, Dr. LeDoux said. But it does not eliminate it.

In further experiments, in which researchers damaged a small region of the rat forebrain, the rats not only did not lose their fear but remained afraid much longer, indicating that the frontal region helps control emotional memories forged in the amygdala and may prevent responses that are no longer useful.

This finding explains why a person who hears a lion's roar in a zoo is not afraid, Dr. LeDoux explained. Input from the frontal area of the brain helps override the fear. But problems with this circuit may underlie phobias, he said. Some people respond with fear to a stimulus such as a lion's roar, even though they know there is no danger. "You can tell phobics all day long, 'This will not hurt you,'" Dr. LeDoux said, "but they don't believe it."

While animal experiments have helped scientists trace exact pathways for fear, the question of how emotions such as joy, sadness, anger or shame are wired in the human brain is more difficult to answer. Psychologists and philosophers have long examined emotions and their impact on behavior, but they have done so by observing what people do and say. Few have ventured into the so-called "black box" of the brain.

But advanced imaging techniques that can look inside the brains of subjects while they talk about feelings and experiences are beginning to lead to a neurobiology of emotions. People with brain damage are particularly revealing in this regard. When specific parts of the brain are damaged, patients may lose the ability to feel emotions, sometimes with disastrous consequences.

Pioneering work in this area is under way at the University of Iowa Medical School, where Dr. Antonio Damasio leads a team that is probing the brains of stroke and accident victims whose personalities have been affected by their injury. Dr. Damasio, a neurologist, recently described his ideas on emotions in a book called *Descartes' Error* published by Grosset/Putnam. The French philosopher Rene Descartes held that morality, reason, language and spirit were held in the lofty brain whereas biology, emotions and animal instincts reside in the body, Dr. Damasio said. The new neurobiology of emotions seeks to overturn this false dichotomy.

Three groups of patients provide clues to how emotions are processed in the brain, Dr. Damasio said.

One group suffers from damage to a small part of the prefrontal lobe, just behind the forehead above the eyes; they invariably undergo a character change. One patient, called Eliot in the medical literature, approached life on a neutral note after his brain injury, Dr. Damasio said.

"In my many hours of conversation with him, I never saw a tinge of sadness, impatience or frustration," Dr. Damasio wrote in his book.

Moreover, Eliot had difficulty making ethical decisions. "I became intrigued with the idea that reduced emotion and feeling might play a role in Eliot's decision-making failures," Dr. Damasio said. Perhaps this area of the brain is involved in personal and social dimensions of reasoning, he said.

The second group of patients suffers from damage to an area on the right side of the brain where sensory signals from the body are processed. Called neglect patients, they exhibit a strange behavior that can be temporary or long-lasting. Although they are paralyzed on the whole left side of their bodies, Dr. Damasio said, when asked if they can tie their shoes or wave their left arm, they say, "Of course I can." Ask them to do it and they say, "O.K., happy to oblige." When they fail to move and the researcher asks why, they say, "Give me time. I'll do it!" Eventually they may say that they don't feel like it now and will do it later.

In neglect patients, the damaged area of the brain is responsible for processing information about the external sense of touch, temperature, pain, internal sense of joint position, state of the limbs, trunk and head, visceral state and pain. This brain area and other regions that it talks to provide the brain with a comprehensive, integrated map of the body's current state of being, he said. While both sides of the brain collect such information for representing extrapersonal space and emotion, the right side is dominant. If the left hemisphere is damaged, neglect does not occur.

A third type of patient has bilateral damage to the amygdala, Dr. Damasio said. Such patients have difficulty recognizing fear in themselves and in others. They would have no qualms about walking down a dangerous street at midnight, he said, and often get themselves into trouble.

The prefrontal lobes, amygdala and right cerebral cortex form a system for reasoning and decision-making in social and personal domains and give rise to emotions and feelings, Dr. Damasio said.

The amygdala and prefrontal regions evaluate a visual stimulus conceptually in terms of earlier experience, Dr. Damasio said, and together they generate a response that is transmitted along two pathways, one to the body proper and one back to the brain.

"The brain gets a double whammy," Dr. Damasio said. It receives a barrage of signals from the body, describing how the body has changed. For instance, fear may be accompanied by the gut contracting, the heart racing and skin turning pale. This is the core of the emotional state that goes to the somatosensory cortex, which is dominant in the right hemisphere.

"At the same time," Dr. Damasio said, "the signal from the brain stem spritzes chemicals and changes the way brain networks operate. The result is that you become aware of your body changes and you also become aware of the fact that something has changed in your mind process."

An emotion or a feeling is a combination of these two things, Dr. Damasio said. The essence of an emotion is the collection of changes in the body state and mind state that are detected by these circuits, he said. A feeling is the experience of those changes.

Primary emotions such as fear and hunger are deeply ingrained in these circuits, Dr. Damasio said. Secondary emotions such as melancholy and shyness are variations on primary emotions and result from experience involving memories and connections between categories of objects and situations.

"The overall function of the brain is to be well informed about what goes on in the rest of the body and about how the body interacts with the external world in order to survive," Dr. Damasio said. "We monitor the background state of our bodies all the time. Hence we usually have an answer to the question, 'How do you feel?'"

A brain has no mind until it can display images internally and manipulate those images in a process called thought, Dr. Damasio said. Thought eventually influences behavior by helping to predict the future, to plan and choose the next action.

It would appear, then, that brain and body co-evolved, "If there had been no body, there would be no brain," Dr. Damasio said. When a brain is deprived of bodily sensations, as in severe spinal cord injuries that result in paralysis from the neck down, the mind is affected, he said. Quadriplegics often describe themselves as being blunted of emotion, Dr. Damasio said.

Dr. Damasio also has an explanation for the state of mind called intuition. To explain why people have emotions and feelings in the absence of strong stimuli, he proposes the idea that the brain can generate signals internally from the amygdala and prefrontal cortex and can send them directly to the somatosensory cortex, in a kind of "as if" emotional loop. Such emotional responses and feelings are less vivid than externally generated ones, Dr. Damasio noted, but they can drive behavior. They are also a factor in intuition—the seemingly mysterious mechanism by which we arrive at a solution without reasoning toward it, he said.

—SANDRA BLAKESLEE, December 1994

New Kind of Memory Found to Preserve Moments of Emotion

A Center for Emotional Memory

The fight-or-flight hormones, adrenaline and noradrenaline, act on pathways in the amygdala to lay down memories of emotional events.

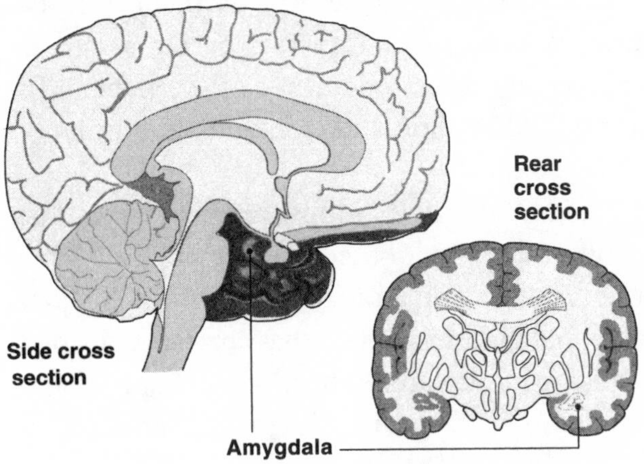

Rear cross section

Side cross section

Amygdala

Source: The CIBA Collection of Medical Illustrations (CIBA-Geigy)

N.Y. Times News Service
Illustration by Jody Emery

DO YOU REMEMBER where you went on your first date? Or the most terrifying scene of the last movie that really frightened you? Or what you were doing when you heard the news that the space shuttle *Challenger* had blown up?

The fact that most people have detailed answers for such questions testifies to the power of emotion-arousing events to sear a lasting impression in memory.

Scientists believe they have now identified the simple but cunning method that makes emotional moments register with such potency: It is the very same alerting system that primes the body to react to life-threatening emergencies by fighting or fleeing.

The "fight or flight" reaction has long been known to physiologists: The heart beats faster, the muscles are readied and the body is primed in the most primitive of survival instincts. These and other distinctive reactions are triggered by the release into the bloodstream of the hormones adrenaline and noradrenaline.

The same two hormones, it now appears, also prime the brain to take very special note in its memory banks of the circumstance that set off the flight-or-fight reaction.

The discovery "suggests that the brain has two memory systems, one for ordinary information and one for emotionally charged information," said Dr. Larry Cahill, a researcher at the Center for the Neurobiology of Learning and Memory at the University of California at Irvine. Dr. Cahill and colleagues published the findings in the journal *Nature*.

The emotional memory system may have evolved because it had great survival value, researchers say, ensuring that animals would vividly remember the events and circumstances most threatening to them.

The findings confirm in humans the relevance of 15 years of research on the neurochemistry of memory with laboratory rats by Dr. James L. McGaugh, director of the Irvine center and a co-author of the paper. His work with animals had implicated adrenaline and noradrenaline in emotional arousal and memory.

"I think it's very exciting," said Dr. Larry R. Squire, a research scientist specializing in memory at the medical school of the University of California at San Diego. "When you study the effects on a rat's brain of having its foot shocked, you don't really know what emotional state that corresponds to in humans—you could argue its analog in humans is sheer panic. But this suggests it's related to more usual emotions, like hearing surprising news, being worried or a little scared."

The new experiment depended on use of a drug known to block the effects of adrenaline and noradrenaline and on seeing if it impaired emotion-laden memories in subjects who have been told a horrifying story. In the study volunteers watched a slide presentation with one of two narratives. In the neutral, rather boring, version a mother and her son go for a walk to visit his father at the hospital where he works; the story describes the bland details of what he saw on the way and while he was there.

But in the upsetting version, the boy is critically injured in a terrible accident on the way and rushed to the hospital, where he is treated for severe bleeding in the brain and a surgical team struggles to re-attach his severed feet.

Before hearing one or another version of the story, half the volunteers received an injection of propanolol, a drug that nullifies the usual effects of adrenaline and noradrenaline by plugging up the receptor sites on the surface of cells that normally respond to the two hormones.

A week later, the volunteers were given a surprise memory test for details of the story. The volunteers who did not get the propanolol remembered more of the upsetting details of the story than the neutral parts, showing that even minor emotional distress enhances memory—a result found in many previous studies.

The key finding was that those volunteers who received the adrenaline-defeating drug were worse at recalling the upsetting details of the story—but not the neutral details—than were those who had no injection. Blocking adrenaline and noradrenaline impaired just the emotional memory of the subjects.

"This is a memory boost system that works in gradations, activating in proportion to the emotional charge," said Dr. Cahill. "We find that it doesn't depend on some intense trauma, but works even when you're just mildly emotionally aroused. But it doesn't activate until there's an emotionally loaded event."

The study is the first to make a definitive bridge to humans from a parallel body of research on emotions and memory in animals. Dr. McGaugh, through a long series of experiments with animals, has pinpointed the amygdala, a pair of walnut-shaped structures that regulate emotion, as the key site where the adrenergic hormones, adrenaline and noradrenaline, affect memory.

"We don't know the precise point of initiation in the brain," said Dr. McGaugh, "but when we get excited about something, a nerve running out of the brain to the adrenals triggers their secretion of adrenaline and noradrenaline." The adrenals are glands that sit on top of the kidneys; when they excrete adrenaline and noradrenaline, the hormones surge through the bloodstream, making the heart beat faster and otherwise priming the body for an emergency.

The adrenaline and noradrenaline appear to activate receptors on the vagus nerve running into the brain. While one job of the vagus nerve is to regulate the heart, it also carries signals to the amygdala. "The noradrenaline activates neurons within the amygdala, which in turn signal other brain regions, presumably cortical areas, to strengthen memory," said Dr. McGaugh. "That's what makes us remember emotionally arousing events so well."

The findings that a minor emotional surge is enough to implant information a bit more firmly in memory might imply, for example, that the anxiety students feel while studying for an exam could itself improve their memory for information—at least to a point. Too much agitation disrupts concentration on what one is trying to read and so interferes with its registering in memory in the first place.

"Psychologists have said for decades that motivation is important for learning," said Dr. McGaugh. "We'd say excitement is important. In my judgment, it would do no harm to make learning more exciting."

Another implication is for preventing trauma in people like rescue workers who know they are about to enter an upsetting situation. The fight-or-flight system seems to play a major role in the troubling and intrusive memories that disturb people with post-traumatic stress disorder. "This suggests it might be possible to mute the formation of symptoms by inactivating this system," said Dr. McGaugh. "People like investigators of airplane crashes could take a propanolol-like drug to prevent traumatic memories."

Still another implication is "a modest alert that some people taking beta-blockers for treatments of heart conditions may find the medication attenuates their memory under emotionally arousing conditions," said Dr. McGaugh, referring to the general name for adrenaline-defeating drugs. Other studies of the effects of beta-blockers on memory have come up with mixed results, but its effects specifically on emotional memory have yet to be studied, said Dr. McGaugh.

The findings also suggest that compounds that enhance, rather than block, the effects of adrenaline and noradrenaline might improve memory in humans, Dr. McGaugh said. That possibility is already supported by work with laboratory animals.

Researchers say they are struck by the elegance of the brain's design for memory. "In evolution, this emotional memory system has obvious adaptive value," said Dr. Cahill. "It's very smart of Mother Nature to build a system that remembers things in proportion to how much it helps you survive—like what to eat and what eats you."

—DANIEL GOLEMAN, October 1994

Brain Study Examines Rare Woman

A BRAIN-DAMAGED WOMAN who cannot recognize fear on other people's faces or feel it herself is helping researchers learn important details about how the human brain is wired.

By observing the woman's behavior, scientists have found that the brain has separate circuits for recognizing facial expressions and facial identities. They have proved that a small brain structure, the amygdala, is essential for experiencing fear and for recognizing other emotions. And they have gained a deeper understanding of brain regions that facilitate social interactions, including making or breaking eye contact.

It has long been known that the amygdala plays a role in emotions and social behavior among animals, said Dr. John Allman, a leading neuroscientist at the California Institute of Technology who is familiar with the research. When a cat's flank is rubbed by a human or another cat, he said, cells in the cat's amygdala start firing.

But the exact role of the human amygdala has been less well understood, Dr. Allman said. If it is involved in social communication, the amygdala probably plays a role in decoding faces, he said, because they convey information about a person's identity, emotions and intentions, which are all critical to social behavior.

The new findings, published in the journal *Nature,* stem from research on a woman with a rare brain disease that dumped calcium deposits in her amygdala—a structure that lies deep within the brain's temporal lobe at the side of the head. Because the brain has two hemispheres, it has two amygdalas; both of the woman's amygdalas were destroyed by the disease, but the rest of her brain was unaffected.

The patient, known as S. M., is being studied at the University of Iowa College of Medicine in Iowa City by a team of neuroscientists, including Dr.

Rolf Adolphs, Dr. Daniel Tranel, Dr. Antonio Damasio and his wife, Dr. Hanna Damasio.

"The woman is 30 years old, nice, cooperative and intelligent," Dr. Antonio Damasio said in a telephone interview. "She has a splendid memory, and when she meets a new person she has no trouble remembering them the next time they meet." Shown the photographs of 19 past acquaintances, he said, S. M. recognized all of them.

The Iowa team has studied a class of brain-damaged patients who cannot recognize familiar faces, including their own or those of their spouses, and cannot learn to recognize new faces. But they can recognize facial expressions like sadness or joy, Dr. Damasio said. Their brain damage is in an area near the middle back of the brain in the higher cortex.

S. M. seems to have the opposite problem, Dr. Damasio said. In an experiment, he said, she was shown photographs of faces with pure expressions of happiness, surprise, fear, anger, disgust or sadness, plus several faces with neutral expressions. She was then asked to describe the emotion on each face with an adjective.

S. M. could not recognize the expression of fear on any face, Dr. Damasio said. She would grope for an adjective but would be stumped; on the other hand, she could usually find correct adjectives for the other facial expressions.

Asked to make an expression of fear on her own face, Dr. Damasio said, S. M. would grimace in a mirror, move her eyebrows around and then exclaim, "I can't do it!"

In a second experiment, S. M. and other subjects, with and without brain damage, were shown faces with a mixture of expressions and asked to rate, on a scale of 1 to 5, how much of any given emotion was present, Dr. Damasio said.

The other subjects could easily pick up the blends of emotions, he said. Their ratings for each expression were remarkably similar. But the woman with no amygdala could recognize only one emotion at a time, with the exception of fear. The complexity of human expressions eluded her.

The experiments have intriguing implications, Dr. Damasio said. How does a person who cannot recognize fear make appropriate decisions? "If someone put a gun to S. M.'s head, she would know intellectually to be afraid but she would not feel afraid as you or I would," he said.

The research shows that people have separate systems for recognizing facial identity and facial expressions. "The normal brain brings them together and is never aware they are separate, like complex music produced by two instruments," Dr. Damasio said, adding that the amygdala is a central part of this circuit.

—SANDRA BLAKESLEE, December 1994

People Haunted by Anxiety Appear to Be Short on a Gene

THEY ARE THE TYPE of people who own a one-sided bed: the wrong side. They are often anxious, grumpy and self-pitying, viewing the past with regret, the present with suspicion and the future with dread. The traditional tag for them is neurotic, but a better word is kvetch.

Now it seems that people who are prone to anxiety and pessimism may have drawn a short stick, genetically speaking. Scientists have discovered a modest but measurable link between anxiety-related behavior and the gene that controls the brain's ability to use an essential neurochemical called serotonin. They have found that individuals who have a slightly abbreviated version of the gene for the serotonin transporter rate higher in negative thoughts and feelings than those with a relatively long rendition of the gene.

The scientists emphasize that the impact of the transporter gene on behavior is quite small, accounting for only about 4 percent of the difference in people's tendency toward neuroticism. They suspect that anywhere from 9 to 14 other genes, as well as many environmental factors that have yet to be sorted out, come into play in making one person anxious, another calm.

"You wouldn't know anything about somebody's personality just by looking at this gene in isolation," said Dr. Dennis L. Murphy of the National Institute of Mental Health. Nevertheless, Dr. Murphy added, "it does seem to be connected in a small way to anxiety."

Dr. Murphy, Dr. Dean H. Hamer of the National Cancer Institute, Dr. Klaus-Peter Lesch of the University of Wurzburg in Germany and their colleagues have published their results in the journal *Science*.

Serotonin, famed as the target of Prozac and other antidepressants, is a so-called neurotransmitter, relaying signals from one brain cell to the next. It helps orchestrate fundamental tasks like eating, sleeping and movement, and also affects mood and thought. The serotonin transporter is a separate molecule that allows nerve cells to respond to the serotonin surrounding them.

The new study was designed to look at garden-variety neuroticism, not the extreme sort of anxiety found in panic disorder and other mental illnesses. Its finding marks the second time that researchers have associated a gene with a normal human personality trait. Earlier this year, scientists announced a link between a taste for novelty and excitement and a gene involved in the activity of dopamine, another of the brain's neurotransmitters.

The work on novelty-seeking has come under fire lately from some scientists, but the new study on neuroticism and the serotonin transporter is considered more persuasive on several counts.

For one thing, the study is quite large. More than 500 people took part in it, the majority of them young, white, male college students.

To determine the degree of the participants' anxiety and neuroticism, the researchers had them fill out personality questionnaires in which they noted the strength with which they agreed or disagreed with statements like "I am not a worrier," or "Frightening thoughts sometimes come into my head."

The scientists also took samples of the participants' blood, from which genetic material was extracted. They found that those with the short type of the transporter gene scored higher on the neuroticism scale than those with the long form, while other personality traits, like extroversion or agreeableness, were not linked to the gene.

But the new study offers more than a statistical association between a gene type and a behavior. Of great importance, that association is buttressed by biochemical evidence. The researchers found that the difference in the two transporter genes occurs in a particular spot, called the promoter, which serves as the gene's on-off switch. In the long version, an extra bit of genetic material is stuck within the promoter. In the short variant, the promoter lacks that DNA insertion.

Often, the presence or absence of a few genetic subunits makes no difference in the performance of a gene. In this case, it does. The scientists determined that the short promoter is relatively weak and that the gene therefore pumps out relatively few copies of the transporter molecule within neurons of the brain. By contrast, the long promoter is robust, allowing the gene to churn forth 1.7 times the number of transporter molecules as its diminutive counterpart produces. The more transporters a nerve cell has at its disposal, the better the cell can react to serotonin.

In other words, the two forms of the gene do not just look different, they act differently, which would explain why the possession of one or the other would, in a minor way, influence a person's temperament.

Because of the functional distinctiveness of the two variants, Dr. David Goldman of the National Institute on Alcohol Abuse and Alcoholism said in an interview, "the data look exciting and convincing."

Dr. Goldman wrote a commentary in *Science* to accompany the new report. He also participated in a study published this month rebutting the connection between dopamine and novelty-seeking and is known jokingly among his colleagues as Dr. No. His praise for the new work, therefore, is seen as another point in its favor.

Still, the transporter gene is but a brief paragraph in the tale of neurosis. After all, with just two promoter variants to choose from, the gene cannot account for the wide range in fretfulness, pessimism and anxiety seen in any given cluster of people. As it turns out, most people—nearly 70 percent—have the less vigorous and more anxiety-provoking version of transporter output.

The skewed distribution of gene formats could reflect natural selection at work, said Dr. Una D. McCann, who studies anxiety disorders at the National Institute of Mental Health. Were it not for a bit of fretful wariness on the part of our ancestors, Dr. McCann said, they might never have survived the odd encounter with a python, a leopard or a rapacious neighbor.

"Anxiety is there for a really good reason," she said. "It's one of the things that is part of our genes because it's protective."

And while feeling tense and peevish may not be much fun, evolution cares nothing for our amusement, but only whether we survive long enough to breed.

—NATALIE ANGIER, November 1996

Provoking a Patient's Worst Fears to Determine the Brain's Role

IT WAS AN ODD REQUEST. The woman, a patient with obsessive-compulsive disorder, was asked to bring two towels from home to a brain imaging laboratory at Massachusetts General Hospital in Boston. One towel was freshly laundered; the other she had used when she washed her hands after going to the bathroom.

For the woman, that used towel was an object of horror and dread. If she held it in her hand, it would trigger an overwhelming train of obsessions about contamination and germs and an almost unbearable urge to wash immediately that if not acted on would set off a state of high anxiety. But despite her loathing, the woman held the soiled towel as she lay still inside the tube of a positron emission tomography (PET) scanner.

The woman was one of several dozen patients with a range of psychiatric problems who, in the interests of science, have volunteered to have their worst symptoms provoked while images are made of their brains. The goal: to capture an image of the perturbations of their brains while they wrestle with their obsessions and compulsions.

The approach is adding a new level of detail to psychiatry's understanding of what goes wrong in the brains of patients when symptoms as diverse as post-traumatic stress, obsessions, phobias and delusions have them in their grip.

"This approach lets us see the brain circuitry that presumably is involved in the symptoms themselves," said Dr. Scott Rauch, a psychiatrist at Harvard University Medical School who conducted the study of the woman with obsessive-compulsive disorder. Dr. Daniel Weinberger, chief of the Clinical Brain Disease Branch at the National Institute of Mental Health in Bethesda, Maryland, said: "It's a long-standing mystery exactly what systems of the brain are active during psychiatric symptoms. If you take a brain scan of a psychiatric patient who is not having symptoms at the time, you don't know if what you see is related to the disorder. But if you evoke the symptoms, you are much surer that what you see physiologically bears a relationship."

The scans of patients with symptoms of obsessive-compulsive disorder showed, for example, increased activity in a series of structures linked to the limbic system, the ancient emotional part of the brain.

Scientists hope that by establishing the unique brain signatures of psychiatric symptoms they will eventually be able to use imaging methods to bring greater precision to diagnosis and treatment. "One day brain imaging may help sort out which patients would benefit from what treatment," said Dr. Rauch.

Dr. Rauch directed the study of eight patients with obsessive-compulsive disorder, as well as a study of seven patients with phobias and another eight who suffer from post-traumatic stress disorder. His group at Massachusetts General Hospital is one of a handful across the country that are carefully provoking psychiatric symptoms in patients so that brain images can be made.

The research strategy is very new. The first published report of the approach, Dr. Rauch's study of obsessive-compulsive patients, appeared last year. Last month researchers at the University of Washington School of Medicine in Seattle published findings in *The American Journal of Psychiatry* on patients who were hyperventilating during panic attacks.

While many previous studies have used brain images of patients with psychiatric disorders, little attention has been paid to the patients' mental state at the moment the images were made. Typically, patients have been asked to lie quietly during the procedure, under the assumption that the brain would be in a "neutral" state.

But as symptoms wax and wane, the images rendered of patients' brains can change drastically. "Simply asking patients to lie quietly fails to control for whether they are happily daydreaming, worried about their taxes or having a panic attack," said Dr. Rauch. "We know now from the wealth of neuroimaging data that the mental state a patient is in is crucial to the image we get."

In the study of patients with obsessive-compulsive disorder, the onset of symptoms was carefully orchestrated with the brain images taken. The woman with the dread of soiled towels, for example, first held the clean towel while she relaxed and inhaled a radioactively tagged form of carbon dioxide, which would highlight her brain activity during the PET scan.

After a 20-minute break while the radioactive markers decayed, she was handed the dirty towel. When she reported that her obsessions were in full sway, she inhaled the radioactive carbon again and a second image was made. Since the woman was touching a towel during both images, the researchers could differentiate between the brain activity associated with touching something and the brain activity associated with obsessive-compulsive disorder symptoms.

"By comparing the two scans, you see exactly which parts of the brain are active during the obsessions," said Dr. Michael Jenike, a colleague of Dr. Rauch.

Psychiatrists met with each of the eight patients in the study to pinpoint exactly how to induce their obsessions during the PET scan. For a man whose obsessions centered on the fear that he might become violent and harm someone, for example, the neutral stimulus was a photograph of his pet dog and the stimulus used to trigger his symptoms was a photo of a serial killer. For another person, the neutral stimulus was a clean dollar bill, and a dollar bill that the patient thought had been used to buy drugs was used to provoke his obsessions about contamination.

During the patients' symptoms, the PET scans showed heightened activity in the paralimbic belt, structures linked to the limbic system. Most active were the insular cortex, a region enfolded deep within the cortex; the posterior orbitofrontal cortex, which lies behind the forehead; the anterior cingulate cortex, which is above and behind the orbitofrontal zone; and the anterior temporal cortex, just in front of the ears.

Because these structures are central sites for anxiety, they are sometimes called the "worry circuit." Many or most of these areas, Dr. Rauch's research shows, have heightened activity during symptoms of any anxiety disorder, whether panic attacks, obsessive-compulsive disorder or phobias. Another area that is active during symptoms of these disorders is the locus ceruleus, part of the brain stem that rouses the body to action by secreting norepinephrine, a brain chemical released when a person is under stress. Dr. Rauch will report these findings in Paris this month at the First International Conference on Functional Mapping of the Human Brain.

In addition to the worry circuit activity, Dr. Rauch has found increased activity in the language and visual areas of the cortex. This activity seems related to how patients experience their obsessions, "seeing with the mind's eye or hearing the mind's voice," said Dr. Rauch.

For a study of simple phobias, Dr. Rauch and his team recruited patients who were terrified of small animals like snakes, rats and tarantulas. In individual consultations, the researchers determined exactly how close each patient could tolerate having the animal placed, so that the patient would be overwhelmed by anxiety and yet could still lie quietly in the brain scanner.

Some patients could tolerate having their hands resting on a plastic box containing the animal. But Dr. Rauch said, "One patient couldn't stand to have it closer than in the adjoining room."

As with the obsessive-compulsive patients, while the feared animals were close by, the most brain activity was in the paralimbic region, especially the anterior temporal cortex and insular cortex, which are associated with a racing heartbeat and escalating blood pressure during anxiety episodes.

There was also increased activity in the tactile cortex, which processes sensations of touch. "When they described their fears," said Dr. Rauch, "the focus was on imagining the feeling of the animal rubbing against them. This was reflected in the heightened activation in the cortical center for the sense of touch, even though they were not actually touching the animals."

In a study of patients with post-traumatic stress disorder, Dr. Rauch and his colleagues conferred with volunteers to write vivid, detailed narratives of the traumatic events that caused the symptoms. A firefighter, for example, helped compose a script based on his experience recovering the charred remains of bodies. These narratives were read to the patients to trigger their trauma symptoms while PET scans were made. For comparison, the patients also wrote narratives that invoked a neutral scene.

Again, while the patients were experiencing symptoms the paralimbic belt showed heightened activity, but mainly in the right hemisphere, which is thought to be active during distressing emotions. Another particularly active structure was the amygdala, an almond-shaped limbic structure at ear level, which is a repository for emotionally charged memories.

The involvement of the amygdala in post-traumatic stress disorder (PTSD) makes sense, said Dr. Rauch, because "PTSD is the vivid evocation of a painful, haunting memory. PTSD victims will tell you that it's not just a memory—during their episodes they feel as though the events were actually recurring all over again."

For many post-traumatic stress patients, that replay is through strong visual images, and Dr. Rauch found heightened activity in the patients' visual cortexes.

Another effect of the trauma was that the language areas of the cortex "turned off," said Dr. Rauch. This, he added, makes sense of a common phenomenon in the disorder in which "patients have difficulty putting words to these emotions, and so have trouble dealing with these experiences."

Other researchers are triggering just a single symptom, hyperventilation, in patients with panic disorder. Researchers asked one group of patients being treated for panic attacks and another group without the disorder to make themselves hyperventilate by breathing rapidly and deeply while images were made of their brains.

"The people with panic disorder, compared to people without the disorder, had a unique brain metabolic response to the hyperventilation," said Dr. Stephen Dager, a psychiatrist at the University of Washington Medical School in Seattle. "We found a perturbation in the insular cortex." He is undertaking further studies to see if other brain regions are especially active during hyperventilation in people with panic disorder.

Dr. John Hsiao, a psychiatrist at the National Institute of Mental Health, has done preliminary brain imaging studies on schizophrenia patients who were taken off clozapine, a medication that had suppressed their psychotic symptoms. "The patients had the full range of schizophrenic symptoms," said Dr. Hsiao. "Some were delusional, some were having auditory hallucinations, though we don't know exactly how acute they were during the scan."

Among the brain areas showing unusual activity while the patients were having symptoms were the medial temporal cortex, which includes limbic structures like the hippocampus and amygdala, as well as the thalamus, the brain's central relay station for sensory data, emotions, movement, thinking and planning. Work by Dr. Nancy Andreasen at the University of Iowa in Iowa City has found structural abnormalities in the thalamus in some patients with schizophrenia.

One of the more distinctive symptoms of schizophrenia, hallucinatory "voices," may be related to activity in a language area, the anterior frontal gyrus. When the patients went back on clozapine, which silenced the voices they heard, brain scans showed that "there was a reversal of the activity" seen in the language area without the medication, said Dr. Hsiao.

—DANIEL GOLEMAN, June 1995

How Brain May Weigh the World with Simple Dopamine System

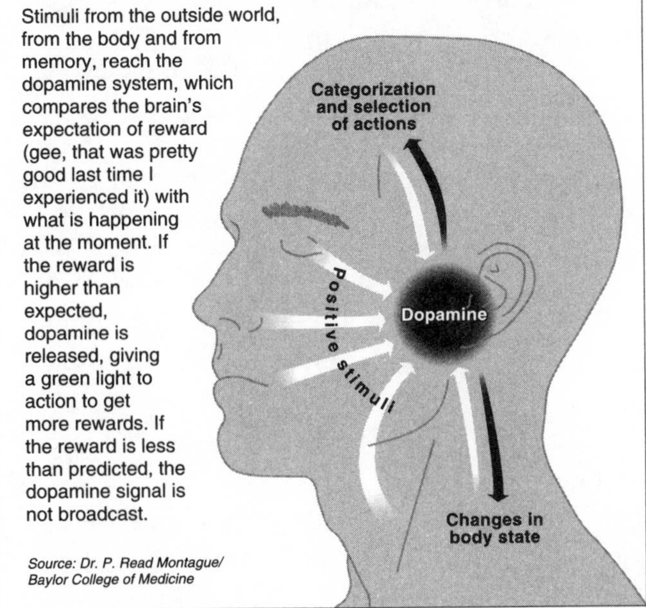

IN THE EVER increasing cascade of new information on brain chemistry and behavior, one substance seems to pop up whenever pleasure is involved. It is even called dopamine, a name that has no connection to the slang "dope" for drugs, but might just as well, considering that dopamine plays a critical role in the way animals and people respond to cocaine, amphetamines, heroin, alcohol and nicotine.

Dopamine has been shown to be a key modulator in an astonishing array of human behaviors. Get too much dopamine in the brain and you hear voices, hallucinate and wrestle with twisted thoughts. Get too little of it and you cannot move. Like Parkinson's patients, you are locked in your body, depressed and joyless.

Dopamine's broad influence is an irresistible lure to scientists. And based on a few new experiments, some researchers think they have the first glimpses of a simple reward system, mediated by specialized dopamine neurons active in all vertebrates, indeed in insects and crustaceans as well. Their ideas are still more hypothesis than theory and have yet to be integrated with the exceedingly complex waxing and waning of other neurotransmitters in the human brain, but the work has drawn the attention and excitement of other prominent researchers.

The dopamine story begins deep in the brain stem with several tiny clumps of cells, together no bigger than a grain of sand. These 100 million or so cells, the only producers of dopamine, form long nerve fibers called axons that reach out to billions of cells in almost every other part of the brain.

Like other neurotransmitters, dopamine allows neurons to "talk" to each other, facilitating the transmission of signals from one brain cell to another.

This is only one small system in an incredibly sophisticated brain. But its size seems to belie its influence. And the reason may be that the brains of humans and other creatures operate on some deceptively simple rules, said Dr. Terrence J. Sejnowski, a neuroscientist at the Howard Hughes Medical Institute and the Salk Institute in La Jolla, California.

An idea recently proposed by Dr. Sejnowski and others is that the dopamine system evaluates rewards—both those that flow from the environment and those conjured up by the mind. When something good happens, the system releases dopamine, which in essence makes the owner of the brain take some action. This account is vastly oversimplified, of course, but Dr. Sejnowski does suggest that the dopamine system works unconsciously and globally, providing guidance for making decisions, when there is not time to think things through.

Recent experiments on bees, monkeys and humans provide the basis of these ideas about the dopamine system, which are by no means proved. They are more in the realm of well-grounded speculation, Dr. Sejnowski said.

Dr. P. Read Montague, a researcher at the Center for Theoretical Neuroscience in the Baylor College of Medicine in Houston, has collaborated with Dr. Sejnowski and others in modeling the way a dopamine-like system works in bees. The bee brain has only one dopamine neuron. (It actually releases octopamine, a close cousin to dopamine that serves the same purpose.) As in other creatures, this neuron sends projections to every nook and cranny of the bee brain.

Bees can find nectar-containing flowers under highly variable lighting conditions, from numerous angles and distances and during different seasons. There could be dozens of yellow flowers, of similar shape and size in a given field, yet only one or two might contain nectar. Bees can very quickly figure out which ones to sample and give the information to other bees.

How do bees do this? At the University of Illinois, Dr. Leslie Real, an experimental psychologist, has built enclosures with artificial flowers spread over the floor. Bees fly around, sampling various amounts of sugar placed in each flower by Dr. Real.

In one experiment, Dr. Real put out blue and yellow flowers. A third of the blues contained high amounts of sugar; the other blues were empty. Among yellow flowers, two thirds contained a small amount of sugar; the other yellows were empty. The experiment was rigged so that the total amount of sugar in the blue and yellow flowers was identical.

But which foraging strategy would the bee prefer? Would it go for high payoff blues? It would get the most sugar for the least amount of work but it would have to tolerate hitting more "empty" flowers. Or would the bee go the "safer" route and sample the yellows, making more work for itself?

In papers published in 1991, Dr. Real reported that bees went to the yellow flowers 85 percent of the time. In seeking reward, they were averse to taking risks.

Fascinated by this finding, Dr. Montague, in collaboration with Dr. Sejnowski and Dr. Peter Dayan of the Massachusetts Institute of Technology's department of brain and cognition, wondered how the bees computed rewards. And so he built a virtual bee inside a computer, with a model of the dopamine system that might explain genuine bee behavior.

"We made a fake bee and let it fly over the blue and yellow flowers" with variable amounts of sugar, Dr. Montague said. Each time a virtual bee landed on a flower, its dopamine neuron was alerted. As in most animals, the dopamine neuron at rest fires signals at a steady, base-line rate. When it is excited, it fires more rapidly. When it is depressed, it ceases firing.

The virtual bee's neuron was designed to give three simple responses. If the amount of sugar was more than expected (based on what the bee knows about similar looking flowers), the neuron would fire vigorously. Lots of dopamine meant lots of reward and instant learning. If the amount of sugar was less than predicted, the neuron would stop firing. Sudden lack of dopamine going to other parts of the brain told the bee to avoid what had just happened. If the amount of sugar was the same, as predicted, the neuron would not increase or decrease its activity. The bee learned nothing new.

This simple prediction model—the dopamine neuron "knows" what has just happened and is waiting to see if the next reward is greater or smaller or the same—offers one explanation for how the bee behavior might arise, Dr. Sejnowski said. When the dopamine neuron encounters an empty flower, it throws the bee brain into an unhappy state. The bee, in fact, cannot stand hitting so many empties. It would rather play it safe and get more numerous, smaller rewards—or no rewards at all—by sticking to the yellow flowers. A paper describing this work was published in the journal *Nature*.

This model is also consistent with what is known about human behavior and the human brain, Dr. Sejnowski said. "Your dopamine system is sitting there, making guesses about the future," Dr. Sejnowski said. "Given the state I'm in now, is it likely I'll be rewarded in the future?"

Here is how Dr. Sejnowski theorizes that the system works. Sensory information flows into the brain from the outside world and from internal representations. What you see, touch, feel, smell, taste, hear and imagine all combine to produce sensory states. These change from moment to moment.

The brain also contains memories and prior experiences about these states. Some are good; you want more. Others are to be avoided.

Both representations—what is happening now and what you know about it from past experience—are funneled to the dopamine system, Dr. Sejnowski said. Then a simple rule is followed. The dopamine system compares the brain's expectation of reward (gee, this was pretty good the last time I experienced it) with what is actually happening at the moment.

If the reward is higher than predicted, dopamine is sent to many parts of the brain, giving a green light to action to get more rewards. If the reward is less than predicted, the dopamine signal is not broadcast and other systems involving avoidance are activated. In both instances, action does not take place until the dopamine system has evaluated the sensory state you are in and detected an error in your prediction about it, Dr. Sejnowski said.

Dopamine cells have no intelligence, Dr. Sejnowski said. They just increase or decrease their firing rates in response to errors in predictions about the world around you. The researchers described this model in *The Journal of Neuroscience*.

Both these papers involve theoretical arguments and computer models. Recent experiments carried out by Dr. Wolfram Schultz and his colleagues at University of Fribourg in Switzerland have shown that dopamine cells in monkeys do indeed behave in just the way theorized. Electrodes were placed in monkey dopamine cells and the animals were given squirts of apple juice paired with a flashing yellow light. At first, the dopamine cells would fire when the animals got the juice or saw the light, since the reward was not expected.

But when they were no longer surprised by the reward, their dopamine cells no longer fired, Dr. Schultz said. Dopamine cells only fire when the prediction is wrong, he said. And they only fire when the stimulus carries a reward.

Dr. Montague supervised an experiment by his graduate student, David Egelman, to test 40 people in an experiment similar to the one done on bees.

People sit in front of a computer screen with two buttons, labeled A and B. Whenever they press one of the buttons, a vertical bar appears on the screen, Dr. Montague said. It represents one dollar. If the whole bar is colored, it means they earn a dollar. If 60 percent of the bar is colored, they earn 60 cents. If 30 percent is colored, they get 30 cents and so forth.

The task is to push either button and to maximize the amount of money you earn over 250 trials, Dr. Montague said. You can take as long as you want and you can scribble notes on strategies you want to try.

But, like the blue and yellow flowers shown to the bee, the game is rigged. Button A gives smaller rewards, but more of them. Button B gives much larger rewards, but they occur infrequently.

Most people behaved just like the bees, Dr. Montague said. They might start by choosing A, which would give them a good return for several tries. But then A might fall to ten cents or five cents. Their dopamine systems could not tolerate this, he said, so they would switch to button B. But B, as rigged, gives many more low than high payoffs, so the player would quickly switch back to A, he said.

Those who keep switching earn far less than those who have more patience in exploring the potential payoffs in button B, Dr. Montague said.

Only two players, both physicists, followed the riskier and more optimal strategy of sticking with B, he said.

This is one reason most people lose money in Las Vegas, Dr. Montague said. Their dopamine system, operating moment to moment in predicting rewards, goes berserk when they are losing and they switch strategies too often. Another reason, of course, is that the odds in every game of chance are in favor of the house, so inevitably, dopamine or no, the overall flow of money in casinos is in, not out. No system of neurotransmitters would allow most people to win at slot machines, for instance. The odds are in the machines themselves, not in the mind.

—SANDRA BLAKESLEE, March 1996

Brain Images of Addiction in Action Show Its Neural Basis

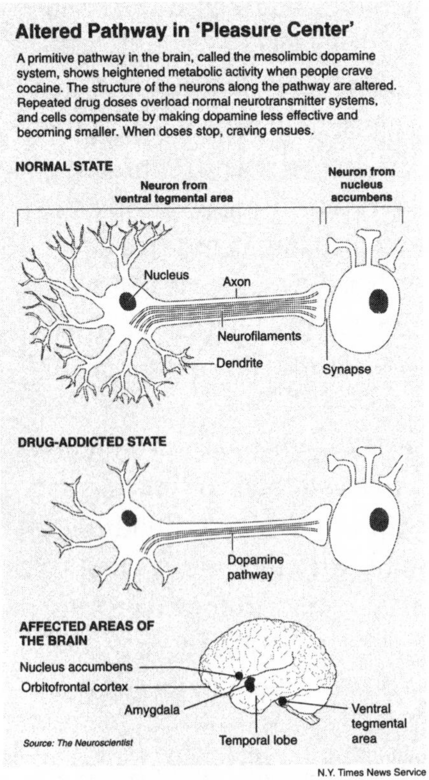

Altered Pathway in 'Pleasure Center'

A primitive pathway in the brain, called the mesolimbic dopamine system, shows heightened metabolic activity when people crave cocaine. The structure of the neurons along the pathway are altered. Repeated drug doses overload normal neurotransmitter systems, and cells compensate by making dopamine less effective and becoming smaller. When doses stop, craving ensues.

Source: The Neuroscientist

N.Y. Times News Service

SOME NEUROSCIENTISTS are experiencing what amounts to a natural high. For the first time, they have captured images of the brains of addicts in the throes of craving for a drug, revealing the neural basis for addiction.

The finding caps a decade or more of intensive brain research seeking the grail of substance abuse, the neurological circuitry that compels addicts to pursue the next fix. And the discovery confirms a number of emerging scientific hunches about the neurology of addiction.

For instance, no matter what the addictive substance is—amphetamines, heroin, alcohol or nicotine—all seem to activate a single circuit for pleasure deep in the most ancient part of the brain. This circuit, for the neurotransmitter dopamine, is the site of the high that addictive drugs bring.

And fine-grained studies of brain cells reveal that repeatedly dosing the brain with addictive drugs is akin to a chemical assault that alters the very structure of the neurons in the circuitry for pleasure. These changes starve brain cells of dopamine, triggering a craving for the addictive drugs that will once again swamp the brain with it.

A drumbeat of recent findings from dozens of scientific laboratories herald these conclusions, which "offer an extraordinary insight into the brain basis of drug addiction," said Dr. Alan I. Leshner, director of the National Institute on Drug Abuse. He added, "There have been a tremendous number of major advances in the last year."

The identity of this brain circuit for addiction is a scientific flashback of sorts, if not a hallucinatory déjà vu: the same brain area was the focus of intense study as long ago as the 1950s, when psychologists routinely implanted electrodes into rats' brains in the region they then called the brain's "reward center." After the rats were trained to push a lever to stimulate this center, they would do nothing else, even forsaking food and water to dose themselves with dollops of rodent bliss—an animal model of addiction. But the specific neural circuitry involved was, at the time, a scientific mystery.

Today that mystery seems to have been solved by using positron emission tomography (PET) scans of the brains of patients being treated for cocaine addiction. Reports from three different laboratories using PET scans show that when addicts feel a craving for a drug, there is a high level of activation in a strip of areas ranging from the amygdala and the anterior cingulate to the tip of both temporal lobes.

This mesolimbic dopamine system, as it is called, shows heightened metabolic activity "when people are in a profound state of craving for cocaine, primed to seek it out and take it," said Dr. Annarose Childress, a neuroscientist at the University of Pennsylvania who did one of the PET studies. The same system seems to be ordinarily in play to provide a sense of pleasure in whatever people find rewarding, like sex or chocolate or a job well done. Dopamine may also be part of a reward system in creatures as different from humans as bees, other researchers have shown.

In Dr. Childress's study, PET scans were done on patients under treatment for cocaine addiction while the patients were being exposed to cues that had made them crave cocaine in the past—like seeing a videotape of people taking cocaine or handling crack pipes or other drug paraphernalia. Drug treatment programs routinely caution patients to avoid such Pavlovian cues, which addicts have learned to associate with the drug high itself, because the cues have long been known to trigger the craving for the drug.

The PET scans showed activation in the mesolimbic dopamine system as the addicts described feeling intense cravings for cocaine.

The mesolimbic dopamine system connects structures high in the brain, especially the orbitofrontal cortex, in the prefrontal area behind the forehead, with the amygdala in the brain's center, and with the nucleus accumbens, a structure that in animal research has proved to be a major site of activity in addiction, although in humans it is about the size of a squished pea, too small to register in PET images. The ultimate source of this dopamine system is the same brain region where psychologists stuck electrodes decades earlier to make rats endlessly stimulate themselves for pleasure, a location called the ventral tegmental area.

These brain areas have emerged in the last several years as hot spots in research on every addictive substance studied and some that create dependency, if not strict addiction. Recently, for instance, Italian researchers reported in the journal *Nature* that the mesolimbic dopamine system was active in nicotine addiction, adding tobacco to a roster that includes heroin, morphine, cocaine, amphetamines, marijuana and alcohol.

In addiction studies with lab animals, a main site of activity is the outer layer of the nucleus accumbens. In humans, a nearby interconnected structure, the amygdala, "is more important in craving," said Dr. George F. Koob, a neuroscientist at the Scripps Institute in San Diego. "If people have a lesion in a section of the amygdala, they no longer link pleasure to its causes—they wouldn't experience a favorite food as enjoyable," he said.

What ties years of brain research on addiction together in a "final bow," Dr. Koob said, is the new finding by Dr. Childress and others that "what lights up during craving is the temporal lobe, particularly the amygdala, where all these pathways converge." Dr. Koob reviewed earlier findings on the dopamine system in the journal *Neuron.*

The various brain pathways he is referring to all have a particular kind of cell that has the D2 dopamine receptor, which is distinct from other dopamine receptors, like those involved in Parkinson's disease. PET images of cocaine patients taken over several weeks after they stop using the drug show a drop in those neuronal activity levels that is consistent with a lessened ability to receive dopamine. Although the degree of this reduction lessens over time, it is evident "even a year and a half after withdrawal," said Dr. Nora Volkow, director of the Division of Nuclear Medicine at

Brookhaven National Laboratory on Long Island. She has also done some of the other recent PET studies.

This pattern of reduced brain activity directly reflects the course of the craving. "The highest risk of relapse for cocaine addicts is during the third and fourth week after they've stopped taking the drug," said Dr. Joseph C. Wu, a psychiatrist at the University of California at Irvine who has made PET images of cocaine addicts that verify the other reports. "You see the lowest levels of activity in the mesolimbic dopamine system during that time."

The brains of addicts are almost back to normal after a year without the drug, though not completely, he said. "If you can stay abstinent for about a year," Dr. Wu said, "you've weathered the periods of greatest vulnerability." Scientists are still debating whether the dopamine cells ever fully return to normal.

The gross patterns of brain activity detected in PET scans represent changes at the microscopic level that are so dramatic that they are akin to the kinds of changes that result from a brain injury, in the view of Dr. Eric J. Nestler, a neuroscientist at the Laboratory of Molecular Psychiatry at Yale University School of Medicine. In an anatomical study of dopamine cells in rats who had become addicted to morphine, Dr. Nestler's team found that the neurons with D2 dopamine receptors had become 25 percent smaller and had lost much of their ability to receive dollops of dopamine from nearby neurons. Their report will be published in *The Proceedings of the National Academy of Sciences.*

The afflicted neurons also underwent a drastic change in their internal dynamics, altering the workings of the so-called second messengers, proteins like cyclic adenosine monophosphate (AMP). After a molecule like dopamine latches on to a receptor on the cell surface, the second messenger acts within the cell to coordinate its response, like the release of neurotransmitters to signal other neurons.

"When you take a drug like cocaine, it floods the neurons with levels of dopamine never seen in nature," Dr. Nestler said. "The addictive drugs have an impact on the dopamine circuitry like a sledgehammer, storming through this pathway with an intensity that never occurs ordinarily. Taking drugs over and over perturbs these systems, and they try to adapt by making the dopamine less effective."

Once the cells adapt this defensive maneuver and become less responsive, the cells are left bereft of normal levels of the neurotransmitter if a person stops taking a substance that floods the mesolimbic systems with dopamine. These changes seem to be the neural engine driving the craving for more of any drug.

"You find the same changes not just with cocaine," Dr. Volkow said, "but also with other addictions, such as to heroin and to alcohol," although each drug affects the dopamine system through distinctive neural routes.

The shift to addiction seems to occur as dopamine deprivation produces chronic unpleasant feelings, depression and a loss of motivation, which leads to the need to take the drug to feel better. "Once these cellular changes occur," Dr. Nestler said, "addicts will take a drug just to feel right, not for a high."

What does all this portend for the treatment of drug addiction? "The research suggests a common biological essence to all addictions," Dr. Leshner, of the National Institute on Drug Abuse, said, "though I don't think we'll ever have a single magic bullet. We might instead one day have neurochemical cocktails that are specific to each addictive drug that would break the cycle of craving."

In the meantime, Dr. Leshner sees a continued role for behavioral treatments of addiction. Approaches that count on people's ability to resist craving, like that of Alcoholics Anonymous, are still the most successful, many studies have found. "If addiction means the brain has changed, then the task is to change the brain back to normal," Dr. Leshner said. "But that doesn't mean treatments have to be biological. Behavioral treatments can change the brain, too."

—Daniel Goleman, August 1993

Does Testosterone Equal Aggression? Maybe Not

TIRED OF HORMONE as cultural myth, as shorthand for swagger and machismo, ferocity and obnoxiousness, the bearskin beneath the three-piece suit?

Do the ubiquitous references to "testosterone poisoning" and "testosterone shock," to "testosterone-fueled heavy metal" and "testosterone-crazed oppressors" make you feel a bit, well, testy? Do you think it unfair to blame one lousy little chemical for war, dictatorships, crime, Genghis Khan, Gunga Din, Sly Stallone, the NRA, the NFL, Stormin' Norman Schwarzkopf and the tendency to interrupt in the middle of a sentence? Ready to give the so-called male hormone a break and retire all testosterone cliches with a single pound of Iron John's drum?

Retire away. As it turns out, testosterone may not be the dread "hormone of aggression" that researchers and the popular imagination have long had it. It may not be the substance that drives men to behave with quintessential guyness, to posture, push, yelp, belch, punch and play air-guitar. If anything, this most freighted of hormones may be a source of very different sensations: calmness, happiness and friendliness, for example.

Reporting at the annual meeting of the Endocrine Society, researchers said that it was a deficiency of testosterone, rather than its excess, that could lead to all the negative behaviors normally associated with the androgen. Studying a group of 54 so-called hypogonadal men, who for a variety of reasons were low in testosterone, Dr. Christina Wang of the University of California at Los Angeles and her colleagues found that before treatment, the men expressed a surprising suite of negative emotions. They did not feel passive or depressed or timid, as standard ideas of testosterone deficiency might predict. Instead, they described feelings of edginess, anger, irritability. Aggression.

When the men were given testosterone replacement therapy and were asked to complete questionnaires about their moods several times over the course of two months of treatment, their general sense of well-being improved markedly. Their anger and agitation decreased, their sense of optimism and friendliness heightened.

"Every parameter we looked at went in the same direction," Dr. Wang said. "The positive mood increased, the negative mood decreased."

Dr. Willis K. Samson, a professor and chairman of physiology at the University of North Dakota School of Medicine, said, "Testosterone has been given such a negative knock. Work like this helps show the up side of this very important male hormone."

The commonly held belief that testosterone produces antisocial behavior "may be a misconception," he added.

Dr. Wang's work is in keeping with similar findings from other laboratories that question how relevant testosterone is to human aggression. Some studies even indicate another, improbable source of aggression: estrogen. Yes, the gal hormone. Other work presented at the meeting showed that when male mice were genetically deprived of their ability to respond to estrogen, they lost a lot of their natural aggressiveness, becoming much less likely to fight with other males or to display the general paranoia exhibited by ordinary male rodents.

Considered together, the new work underscores how primitive is scientists' understanding of the effects of hormones on human and even animal behavior. Testosterone was first isolated nearly half a century ago, yet its influence on the brain and behavior remains largely a matter of creative speculation. "It's more an art form than a science form," particularly when it comes to human research, said Dr. William J. Bremner, an endocrinologist at the Seattle Veterans Affairs Medical Center and the University of Washington.

What is not open to debate is that to call testosterone or any of the related androgens "male" hormones, or to call estrogens "female" hormones is a misnomer: Both sexes possess some amounts of each other's hormones. But they differ considerably in the relative amount circulating through the blood.

In the case of testosterone, women on average have about 40 nanograms of the hormone per deciliter of blood, and a score of 120 would rank as extremely high (and very likely result in hirsutism, acne and other problems of excess androgens). Men, by contrast, normally have 300 to 1,000 nanograms of testosterone per deciliter, or 10 times a woman's concentration.

This discrepancy explains in part why testosterone has been tagged as the hormone of aggression. Men have so much more of it than women—more of it than women have of estrogen relative to males. Men also commit the vast majority of violent crimes and get into many more accidents than do women. There has been other evidence implicating testosterone as the bearer of brutality. When scientists inject laboratory animals with testosterone, male and female animals alike respond with more aggressive behavior, becoming more likely to attack intruders or to begin mounting anything that moves.

Moreover, some studies of prison populations, wife beaters or other groups considered hostile and foul-tempered, suggest that aggressive men have comparatively high testosterone levels. And then there are those football players or weightlifters who take super-high doses of anabolic steroids—synthetic androgens—in an attempt to build strength and muscle mass. Often, such men report feeling prepared to go out and puree their opponents.

But scientists are beginning to question the relevance of animal behavioral studies to human emotions and even to wonder what subsidiary effects the testosterone injections could be having on laboratory animals to explain their increased aggression. As for the human studies, they are contradictory and open to various interpretations. For example, stress can affect hormone levels in ways that are only now being mapped out, and prisoners are likely to be under extremes of stress.

Dr. Wang notes that while competitive athletes may report feeling pugilistic on anabolic steroids, "this is a highly select group of subjects with a particular psychological makeup." And the drugs they take are not native testosterone, but usually a synthesized mix of androgens, with who knows what effect coming from each.

So far, there have been no good, controlled studies seeking to evaluate the effect of giving excess testosterone to androgenically normal men. In lieu of such studies are recent experiments comparing the mood states of hypogonadal men before and after treatment, like Dr. Wang's work.

One missing element of this report, however, is a conventional control group—seeing what happens to men with low testosterone if they are given a dummy medication rather than real testosterone. Do they, too, feel happier and friendlier? Such an experiment would be unethical, said Dr. Wang, because hypogonadal men who visit the clinic are ill and need treatment to restore muscle and bone mass and healthy cholesterol levels. Denying them testosterone would be like denying a diabetic insulin.

Offering a partial explanation, Dr. Bremner and his colleagues reported in *The Journal of Clinical and Endocrinological Metabolism* their results of seeing what happens to healthy men when they are artificially and temporarily brought to a state of low testosterone. Some of the subjects were immediately given testosterone replacement, while others received dummy medication.

The study was mainly intended to look at the effects of testosterone on libido, but the researchers noted that the men with a drug-induced state of hypogonadism reported increased levels of aggression. "Perhaps it made them uncomfortable or unhappy in a variety of ways, and that unhappiness registered itself as a high score on the aggression scale," Dr. Bremner said.

What Dr. Bremner's study and other reports found that does conform to stereotypical notions of the male hormone is that testosterone is profoundly important to a man's sex drive, though not to his mechanical abilities in bed. Hypogonadal men report a sharp drop in sexual interest, which testosterone replacement quickly restores. The androgen may also play a role in female sexuality, and a growing number of menopausal women are asking that testosterone be added to their hormone replacement regimen to restore a lackluster libido. But the data linking sex drive and testosterone in women are fiercely debated.

Testosterone therapy also appears to give men and women more energy, vim, the desire to leap out of bed in the morning and embrace the demands of the day with can-do concentration. That zestiness is not the same as aggression, which if anything is often accompanied by poor concentration and underlying malaise, researchers said.

If testosterone is not the demonic potion of legend, its yangian counterpart, estrogen, may not be so innocent. Reporting at the annual meeting of the American Pediatric Society, Dr. Jordan W. Finkelstein, Dr. Howard Kulin and their colleagues at Pennsylvania State University said that they compared the effects of giving estrogen therapy to girls who suffered from delayed onset of puberty with that of giving testosterone to boys who likewise were late in sexually maturing. The girls showed earlier and larger increases in aggression than did the boys, until the boys received the last and highest dose of testosterone.

The researchers propose that for both sexes, the cause of the teenage spike in aggressive and very likely insolent behavior is estrogen. As scientists only lately are beginning to appreciate, most of the effect of testosterone on the brain is paradoxically estrogenic in nature. That is because the brain is rich in the enzyme aromatase, which converts testosterone into estrogen. The newly transformed hormone then acts on the nerve cells of the brain through estrogen receptors, proteins designed specifically to link up with it.

A male's brain also has some receptors for testosterone, but they are far fewer in number or distribution, and the converting enzyme aromatase does not leave much testosterone around to hook up with these androgen receptors anyway.

Thus, in both boys and girls, as they reach adolescence and their respective sex hormones surge, the influence of either hormone on the brain and behavior probably works its dark art as estrogen. In the Pennsylvania study, the girls may have had a jump on aggressive behavior over the boys because they were given direct injections of estrogen and therefore their brains did not need to go through the work of converting testosterone to estrogen.

The centrality of the brain's estrogen receptors to aggressive behavior was highlighted by a new study of receptor-deficient mice, presented at the endocrine meeting. Dr. Donald W. Pfaff of Rockefeller University in New York, a research associate, Sonoko Ogawa, and Dr. Kenneth S. Korach of the National Institutes of Health have analyzed male mice genetically altered so that they lack nearly all estrogen receptors.

Testing the male mice in a variety of circumstances, the researchers determined that they were unusual in many ways. Normal male mice do not tend to wander across open fields as females do, but prefer to skulk along borders; males without estrogen receptors generally took the female attitude and freely walked where they pleased. Ordinary males respond to intruders in their territory with violent attacks, chasing, biting and generally seeking to harm the interloper. These males react to newcomers tepidly if at all, perhaps nipping if the animal comes too close, but never attacking the stranger outright.

Significantly, the altered males still have all their androgen receptors intact. It is only the ability of their brain to respond to estrogen that is defective.

The researchers have yet to report on the behavior of female mice lacking estrogen receptors. Those results will probably break a few paradigms of their own.

Until then, perhaps it is time for a new hormonal cliche to explain aggression. How about this: The estrogen was so thick you couldn't beat it down with a rolling pin.

—NATALIE ANGIER, June 1995

The Brain Manages Happiness and Sadness in Different Centers

THE ESSENCE OF EMOTION—the rapture of happiness, the numbness of depression, the angst of anxiety—is as evanescent as a spring rainbow. It is hard enough for a poet to capture, let alone a neuroscientist.

Now brain researchers, in their own fashion, have begun to do so. A major result emerging from the new research is that the brain does not have just a single emotional center, as has long been believed, but that different emotions involve different structures. Another is that the brains of men and women seem to generate certain emotions with different patterns of activity.

The new advances are made possible by fast imaging methods that allow researchers to take snapshots of the brain in action. The snapshots are short enough that they roughly parallel the duration of an emotion, however fleeting. They have already resulted in a radical redrawing of the neurological map for emotion, showing regions of emotional activity both in and beyond the limbic system, a ring of structures around the brain stem, which for 50 years was considered the brain's emotional center.

One surprising result of the remapping is that emotional opposites, like happiness and sadness, are not registered that way in the brain, but rather entail quite independent patterns of activity, according to a report in *The American Journal of Psychiatry*.

"It's because happiness and sadness involve separate brain areas that we can have bittersweet moments, like when a child is leaving home for college and you're sad, but happy, too," said Dr. Mark George, a psychiatrist and neurologist at the National Institute of Mental Health in Bethesda, Maryland, and the lead author of the report.

When a woman feels sad, Dr. George discovered with a brain imaging method known as positron emission tomography (PET), her brain shows increased activity in the structures of the limbic system near the face and more activity in the left prefrontal cortex than in the right. His studies were conducted in women to avoid the confounding difficulty of possible differences between the sexes.

When his 11 subjects felt happy, the characteristic pattern was a decrease of activity in the regions of the cerebral cortex that are committed to forethought and planning. These regions are in the temporal parietal area of the cortex, located just over and a bit behind the ears, and the right prefrontal lobe, just behind the forehead. "Those neocortical regions are used in complex planning—it's interesting these shut down in happiness," Dr. George said.

The cue for sadness was to ask the subjects to recall personal events in their lives, such as deaths and funerals, or to look at a picture of a sad face. For happiness, the cue was to remember joyous times such as births and weddings, or to look at happy faces.

In earlier research, Dr. George found that the neocortical areas become even less active when volunteers received injections of morphine or cocaine. "There seems to be a continuum in brain activity in the same regions from transient happiness to ecstasy," he said.

Another key change was in the amygdala, the pair of almond-shaped structures in the limbic system. The amygdala area "activates during sadness," Dr. George said. But the structures change only slightly when a person is happy. "The left amygdala seems to shut down a bit, while the right goes up," he said.

Such findings are mapping out a new neuroanatomical atlas for the emotions, one that eventually may give psychiatrists new guides to treating mental illness. "The brain mechanisms of emotional change are perhaps the most central question in psychiatry," said Dr. Robert Robinson, chief of psychiatry at the University of Iowa Hospitals and Clinics in Iowa City.

Many serious psychiatric disorders, such as depression and panic attacks, are extremes of ordinary emotion. Studies of anxiety, for instance, show that the brain regions that are most active while people are anxious are even more active during panic attacks. Locating the primary sites of various emotions represents a major step toward understanding what is going wrong when these sites become overactive.

The recent findings on sadness offer a new twist: Brain areas involved in ordinary sadness almost completely shut down when a person is clinically depressed. "Sadness and depression seem to involve the same brain region, the left prefrontal cortex, in different ways," said Dr. George. "It gets more active during ordinary sadness, but shuts down in people with clinical depression. Perhaps the left prefrontal cortex somehow burns itself out when sadness persists for several months."

Many people with severe depression no longer feel sadness or any other emotion. "They're emotionally numb," said Dr. George.

He has also studied the locations of happiness and sadness in men, though these studies have not been published. He has found that the processing of emotion is yet another aspect in which the brains of men and women apparently differ. "When they are sad, women activate the anterior limbic system much more than do men," said Dr. George. "At the same time, women seem to experience a more profound sadness than do men. It makes me wonder if this might be related to why women have twice the risk of depression as do men."

In a study still in progress, Dr. George is mapping anger and anxiety. "Other work on anxiety implicates the right temporal area of the cerebral cortex, and our findings seem to support that," Dr. George said.

For anger, a main area of increased activity appears to be the anterior septum, which is in the center of the brain. "In cats, if you stimulate this area with an electrode the cat lashes out in rage at anything nearby," said Dr. George.

Before imaging, neuroscientists' principal method of mapping the sites of emotions in the brain rested on analyzing what was missing in patients who had had brain injuries or strokes. The technique is like drawing a diagram of a house's electrical wiring by pulling out the fuses one by one.

But the brain imaging techniques are still far from perfect. PET scans require subjects to be injected with a mildly radioactive chemical and make images that are averaged from multiple readings instead of from a single scan. And a serious problem with the magnetic resonance imaging technique is that a patient must lie in a metal cylinder, an experience that has been likened to being trapped in a coffin. "While you're in those machines it's very difficult to have reactions other than those to the machine itself," said Dr. Paul Ekman, director of the Human Interaction Laboratory at the

University of California at San Francisco, who recently spent three hours in such a machine.

To overcome the problem, researchers have gotten their subjects into the desired emotional states with such prods as a scene from the film *The Godfather* in which a decapitated horse's head is found in a bed, a joyous scene from *On Golden Pond* in which Jane Fonda dances with her father, Henry Fonda, and even—to elicit fear in people with a snake phobia—a live (but friendly) python perched atop the doughnut-like PET scan equipment that surrounds the head. Patients having magnetic resonance imaging can watch films through a kind of periscope.

Dr. Ekman and colleagues plan to study how emotions are evoked by seeing which brain areas light up in response to stimuli like remembering an upsetting event, seeing an image of a sad face or putting one's facial muscles in the configuration typical of various emotions.

The investigation is of crucial importance for researchers, since the means they employ to evoke an emotion while capturing brain images may itself affect the image they get.

A team led by Dr. Richard Lane at the University of Arizona in Tucson compared the brain areas involved when people either watched film clips that evoked happiness or sadness, or called to mind happy or sad moments. In all cases there was heightened activity in the thalamus and the prefrontal cortex, suggesting a role for these regions in each of these emotions, no matter how it was evoked.

"The prefrontal cortex monitors a person's emotional state, no matter what it is, to generate an appropriate response," said Dr. Lane, "while the thalamus participates in how that response is executed."

During the film—not during emotional memories—two key areas of the limbic system were active: the amygdala and the hippocampus, suggesting that these structures are involved in evaluating whether a situation is of emotional import, Dr. Lane said.

On the other hand, during the recall of sad or happy events there was more activity in the anterior insular region, an area of the cortex with strong connections to the limbic system, implying a special role for this area in emotional memories. "The anterior insular region seems to be involved in investing thoughts or memories with emotional significance," said Dr. Lane.

That same region is also active during anticipatory anxiety, such as when someone is waiting to receive a mild electric shock, according to new findings by Dr. Eric Reiman, a psychologist at the University of Arizona in Phoenix and a colleague of Dr. Lane's.

"The vast majority of findings until now about the brain's emotional regions has been based on emotions in laboratory animals, all of which are induced by external means, such as shocks," said Dr. Reiman. "But the results with people who are evoking emotions through memories are showing that a very different set of brain areas are active than had been thought from the animal research."

The exploration of the brain's topography for emotions through imaging is at its earliest stages and, like any forays into new terrain, may produce distorted maps, researchers warn.

"We need to be cautious about interpreting these findings on the regions involved in emotions," said Dr. Richard Davidson, a psychologist at the University of Wisconsin and a participant in some of the research. "We're just beginning to work out the technical difficulties in capturing neuroanatomical images from people during something so private and fleeting as an emotion."

—DANIEL GOLEMAN, March 1995

3

THE MAKING OF MEMORIES

Ask someone to describe a tiger and from their memory banks they will summon the snarl, the stripes, the stealth, the fearful symmetry. But how and where is all this information stored, ready for retrieval at will?

One feature of the brain's memory management is that it seems to store memories in distributed form. There is no nerve cell equivalent of an encyclopedia entry, containing everything known about tigers. Instead, the information is stored all over the brain, probably near the places where it was first received, with visual information in the optic cortex, which processes sight, auditory information in the auditory cortex and so forth.

Study of patients with specific areas of cerebral damage suggests that the brain has an indexing system for noting objects in general broad categories, like plants, animals or household objects. Presumably it is via this indexing system that relevant information can be retrieved.

The brain's system of storing information is still very much a mystery but recent research has elucidated the central role of an organ called the hippocampus in laying down long-term memories.

Short-term or working memories are formed in prefrontal lobes of the cerebral cortex and then transferred to the hippocampus. They disappear from the hippocampus in a matter of weeks, but during this time the hippocampus forwards them on in some manner to the cortex. Patients with damage to the hippocampus can recall information from before the damage occurred but cannot form new memories.

Progress is also being made in understanding the changes that occur in nerve cells as memories are consolidated.

"Hole" in Tumor Patient's Memory Reveals Brain's Odd Filing System

A SMALL DISCRETE HOLE appeared in the memory of the 70-year-old librarian.

She had had a tumor, but all her faculties seemed otherwise intact. So it was a surprise to discover that she could no longer look at a picture of an elephant and say elephant. Instead, she drew a blank.

Researchers at Johns Hopkins described the curious case in *Nature* magazine this month, saying the woman had lost the ability to name animals or to describe physical attributes, like the number of legs on a dog.

She knew that apples are red and celery is green, but could not name the color of elephants, polar bears or cardinals. Shown a picture of a dog, she could not say what it was. She knew it was a pet, knew it was not food, but still could not name it.

The importance of this and similar cases is that they illuminate the brain's system for organizing knowledge.

For a century or so, psychologists pondering the brain's memory-handling system have suspected that the brain had some system of putting information in categories, with a separate pigeonhole for categories like dogs, plants or numbers, each in a separate network of cells.

The evidence for this thesis rested on patients with minor brain damage whose only problem was that they could not remember or name highly specific categories.

For example, the psychologist J. M. Nielson reported in 1936 on patients who, unknown to themselves, could not name certain inanimate objects. Other patients had no trouble with inanimate objects but could not name certain living things.

"So the question is, what does the brain use to divvy up its knowledge?" said Dr. Barry Gordon, a cognitive scientist at Johns Hopkins Medical Institutions. The Dewey Decimal System, perhaps?

While some categories seem sensible enough, like food versus non-food items or nouns versus verbs, others are a little strange. For example, there was the case of a lawyer who after a bout of encephalitis could not name small household objects like scissors or irons, but had no problem citing large household objects like refrigerators or stoves.

Researchers have seen isolated cases like this over the last century, but in the past decade a new surge of information has led to the identification of 15 to 20 different categories that appear to be the brain's own natural categories for knowledge.

Among them are plants, animals, body parts, colors, numbers, letters, nouns, verbs, proper names, faces, facial expressions and a single category including food, fruits and vegetables. And there are likely to be many more.

Besides their interesting patient, the Johns Hopkins group has also made an important new finding. The researchers have been able for the first time to paralyze their patient's access to a specific category of memories, in this case the ability to compare sizes of things, and then to restore the ability.

The experiment began with a 39-year-old woman who has had epilepsy since she was 19. The storms of electrical activity she experienced as seizures were centered on the left side of her brain, near the temple. Drugs failed to control the seizures, and she was taken to the hospital for possible surgery to cut out the portion of her brain that was the focus of seizures.

To this end, her skull was opened, and the leathery covering of the brain called the dura was opened. To determine how much of the brain should be cut out, a grid of small electrodes to monitor electrical activity was inserted at various spots in the brain's surface to pinpoint the center of the seizures.

As the patient was being observed over a period of 12 days, she was given a standard battery of tests on her mental agility, one of which showed something odd: When the probing currents of the electrodes were turned on, she was unable to tell the difference in sizes of things.

Asked which was bigger, a tree or an ant, she could not say. She was also unable to answer more subtle size questions; for example, whether an apple or a lemon is bigger.

Dr. Gordon and his colleagues, Dr. John Hart and Dr. Ronald P. Lesser, then explored further her ability to make size discriminations.

They discovered that the effect was created by a pair of electrodes on the brain just above the ear and that the condition was reversible when the current was turned off. Surprisingly, the failure to make size comparisons was only verbal and not visual. When shown pictures of two objects she could point to the larger.

The woman was able to discriminate well in regard to other attributes, like shape, color, orientation and texture. She was also able to tell size differences if she had to tell the difference from looking at pictures only.

So, Dr. Hart explained, the loss was verbal only, not visual. The group said this, combined with the information from the librarian who could not name animals, demonstrated that among the other storage strategies, the brain has separate verbal and visual stores, in addition to the cognitive categories like animals or body parts.

The Hopkins group is continuing to monitor patients before brain surgery to see if any other unusual losses of knowledge occur when electrodes are turned on in different areas of the brain.

Dr. Gordon said he found it odd that the brain respected semantic categories and even odder the nature of some of the categories.

"The brain doesn't let information become blooming, buzzing confusion," he said, "It divvies things up and puts them in bins. And these bins may be different from what you would expect. The brain is not necessarily built the way your mind thinks it is."

There are apparently different brain systems to deal with categories of knowledge, neurons actually firing according to abstract ideas.

"And some of these are pretty weird," he said. "If I were designing a brain, I would not have put the sizes of things in a separate category from the things themselves. Or the small household objects as a category—what can that be about?"

The experiments make another point about thinking. "We have multiple representations of the same things in our brains," Dr. Gordon said. "Intuitively, we think that an apple is one notion, one idea. But it's not. An apple's properties are split among many different representations."

Its color, its shape, its size, its qualities as a food, and probably many other features, are each stored separately. Sometimes information is duplicated; for example, the brain's visual representation of an apple and its verbal representation are stored separately.

At key times, they are all activated and become available to consciousness as a bundle, and thus seem to be a single impression with many facets.

Eventually, the work at Johns Hopkins may offer practical information about sensitive locations in the brain that should be avoided during surgery. In addition, because some information is duplicated in the brain, for example, visually and verbally, it may be possible to retrain patients who have brain damage in one system by using the one still intact.

—PHILIP J. HILTS, September 1992

Brain's Memory System Comes into Focus

ONE OF THE DEEPER MYSTERIES of the mind—how people channel the rushing waters of perception and turn them into pools of accessible knowledge—is beginning to be solved.

Since the mid-1980s, with the marriage of sophisticated brain scanning techniques and computer wizardry, it has become possible for the first time to observe directly how the brain manipulates information and to resolve centuries-old disputes about what it means to know something.

"We have really made fantastic progress, and there is now general agreement on the basics," said Dr. Stephen Kosslyn, a neuroscientist at Harvard University. "It has all come together just in the past few years, due to two technical developments—computers and modern brain scanning."

The two work synergistically. "The computers gave us a way to think about information processing in a precise way, avoiding the pitfalls of phrenology but learning how processing is localized, and the brain scans give us a way to test our ideas directly for the first time," he said. "It was the confluence of having a way to think about information processing and the means to test our ideas that broke the field open."

In that brief time, researchers have located the regions where different concepts are processed and where the components of concepts are stored, and they have assembled a number of clues about how they are joined together in thought and consciousness.

At the University of Iowa Medical Center in Iowa City, Dr. Hanna Damasio, a neuroscientist specializing in scanning techniques, sits before a computer screen showing a three-dimensional landscape of the folds and fissures of a thinking human brain. Along the top middle of this image, in a long crevice separating the front from the back of the brain, are ominous dark spots, the marks of brain damage.

Small spots of brain damage may arise from many kinds of injury, from falls to infections to strokes. These damaged areas may ultimately result in disabilities because the brain uses many small, localized processing engines to handle perception, memory and thinking.

On Dr. Damasio's computer, she has maps of the brains of nine patients, each of whom has the same odd disability: They are unable to use ordinary verbs.

Shown a picture of someone working with a broom and asked what the person is doing, they cannot say "sweeping." They may say that the person has a broom, that there is dirt, that the person wants the floor to be clean. If pressed hard, they might invent a word, saying the person is "brooming."

If Dr. Damasio's theory holds, these people will all be found to have damage in the same area of the brain. She taps a key, and the nine images come together in a composite: All of the damaged spots cluster a few inches behind the forehead, and all overlap in a tiny region there. This is near the brain's motor area, where human action is worked out, and it makes sense that words of action might also be processed here.

When Dr. Damasio taps a few more keys, the image on the screen changes and another composite brain is revealed. On this one, the dark spots, also assembled from several different patients, are arrayed in the temporal lobe, a stripe along the lower left side of the brain, fore and aft of the ear.

These patients have a different problem: They can use verbs as readily as anyone, but they have trouble naming objects. When shown a picture of a common animal, for example, one person cannot think of the word "cat." Another person has no trouble with animals, but cannot come up with the names of ordinary tools.

The array of "category deficits" in human thinking that arise from such damage to the brain is astonishing—otherwise completely normal people who simply cannot call animals by their correct names; people who can write words but not numbers; people who cannot recognize small tools and utensils, but have no trouble with large machinery or natural objects; people who cannot recognize the faces of their loved ones.

The range extends from virtually unnoticeable gaps to the near-complete inability to comprehend language, or even the loss of the entire "self."

When studied in detail, across many patients and many different problems of knowing, the various disabilities may give the best picture so far of how the human mind organizes and manipulates all its information, said Dr. Barry Gordon of Johns Hopkins University Medical Center in Baltimore.

Damage to a particular part of the back and sides of the brain can make a person fail to recognize a violin as a musical instrument or his pet poodle as a dog, or any number of other individuals as belonging to particular categories.

One patient, known to the researchers as P. S. D., can recognize categories of natural objects only about 30 percent of the time or less. But his recognition of man-made tools remains good.

"In other words, the same attentive and intelligent patient who quickly recognizes a saw, a screwdriver, a shovel or an electric shaver, agonizes over the recognition of most animals and food items, venturing hesitatingly that perhaps they are some kind of animal or plant," wrote Dr. Hanna Damasio and her husband Dr. Antonio Damasio in one of their many articles on mental categories. P. S. D. could name only 30 percent of musical instruments by sight, but by sound got them all quickly.

While cases like this have been observed over the last century, in the last decade researchers have tentatively identified roughly 20 categories that the brain seems to use to organize knowledge. Among them are fruits and vegetables (one category), plants, animals, body parts, colors, numbers, letters, nouns, verbs, proper names, faces, facial expressions, several different emotions and several different features of sound. There are likely to be many more.

But what kind of information is stored in the brain's category areas? The new scanning machines and other lines of evidence point to a remarkable answer: The items in the animals category, say, are not multimedia sounds and images of zebras and other animals, but rather a set of indexes that draw such information from other parts of the brain.

Each specific realization is achieved by processing in a different area, from recognition of fur to moral judgments. Through each portal of the "senses"—hearing, smell, emotion, balance, touch—the computations flow simultaneously.

The highest level of knowledge, at the end of the line of processing and thinking, are features such as the ability to know specific individuals—one's spouse, pets, cars.

Scientists differ somewhat on the details of how these systems work, but there is general agreement about the basic idea, says Dr. Nina Dronkers of the University of California at Davis. "This is all quite new, within the past 10 years," she said. "As we learn more about the possibility of computation in distributed networks, it seems clearer and clearer the knowledge is organized in bits distributed widely across the cortex."

She noted, for example, that when people think of a tree, "we know the sound it makes in the wind, the look of the trunk, the shape of a leaf.

"Each of these may come from a different place," she added. "Then, there are the arbitrary, linguistic features which we have learned to associate with tree—the sounds for t, r and e, put together.

"The idea that this knowledge is distributed is gaining wide acceptance," she said. "The question now is, where are these bits of knowledge? Are they kept in the areas close to where the senses bring them into the brain, or are they closer to where the information is handled—say in the temporal cortex?"

Dr. Hanna Damasio is the image specialist and Dr. Antonio Damasio is the theorist. He is the author of the notion of "convergence zones," or indexes that draw information from elsewhere in the brain, and has written a book, *Descartes' Error,* discussing the issues.

In his version of how things might work, these indexes do not contain the memories themselves, only the instructions on how to rekindle many related features and memories associated with them.

He suggests that the little engines of processing that handle categories, such as verbs or nouns, tools or animals, are actually indexes or switching centers.

That is, the memory of an animal is not stored whole, but remains where it was when it was first perceived—as features such as legs, general shapes, colors, fur or skin, and so on. These bits of memory are stored near where the sensory impressions first enter the brain—for the visual system this begins in the back of the brain and wraps around to the temporal lobes on the sides.

After the features are combined and recognized, an "index" of these features is formed and stored. Thus, when someone mentions "zebra," the index will light up and simultaneously signal each of the animal's features stored at the back of the brain simultaneously, and then link them to other associated categories, such as horses or African plains.

Under "cats," for instance, might be indicators of where in the brain to activate information about four-leggedness, furriness, cat colors, pointed ears, the distinctive cat-shaped eyes and nose, dogs as enemies, and eating birds and mice, as well as images from African wildlife films, scenes of domestic cat bliss, and so on.

One important feature of this model is that it explains a major fact of the brain that did not fit well in any theory before, said Dr. Larry R. Squire, a neuropsychologist at the University of California at San Diego. As information is processed, moving from station to station in the brain, each station makes many connections reaching back to the earlier levels of processing.

By making these connections, the system can work in reverse. If a feature is triggered in the back of the brain—say a coat of black and white stripes—it can quickly light up the index area, which in turn brings up all the other parts of the zebra, as well as the association with the horse and recollections of a trip to Africa.

Data are gradually coming in which support this idea, but, Dr. Squire says, it is too early to be confirmed.

In the Damasios' model, the memories themselves are stored in bits and pieces near the gateways of the senses, but the higher level binding of them into categories takes place farther along in the processing stream.

For the scattered elements to be bound together and triggered as one experience, and at the same time for one feature—ears, for example—to immediately resurrect the entire image of the horse, the early processors and the later processors must be connected.

Other researchers think that perhaps clots of memories themselves are not stored at the sensory gateways of the brain, but near the sites of indexing.

Dr. Dronkers of the University of California suggests that it might be easier for the features to be bound together if they were present at the site of the indexing. A person may take longer to dredge up the answer to a question about cats than one about dogs if the index and the requisite memory of dogs are nearer each other.

One thing the new ideas suggest is that individuality is the hallmark of all knowledge. Categories are built up out of a person's experience of the world. Some will be similar from person to person, but many will be different—entomologists may devote larger tracts of processing to the categories of beetles, while lexicographers fill their neuronal assemblies with information on words. The person who lives in the desert will have a different experience of trees than the city dweller, and the blind person will have a radically different set of experiences from both of them.

Some scientists suggest that this penetration of categories of knowledge may be just a beginning. Dr. Frank Kiel, a cognitive psychologist at Cornell University, said that while the work is an exciting beginning, it is still tied to its perceptual roots. There may be some mental processors that impose order from above, rather than just guide the stream of data from the senses into appropriate categories.

He believes that the brain may come pre-wired with some ideas—for example, a sense of "social" objects in the world versus ordinary physical ones. People think differently about each other than they do about furniture.

There is a lot of work to be done, Dr. Kiel notes, and ideas are likely to change rapidly as scientists learn what the brain has in mind.

—PHILIP J. HILTS, May 1995

In Research Scans, Telltale Signs Sort False Memories from True

THE TRUTH OF MEMORY is infinitely hard to establish, as psychotherapists, lawyers and scientists know all too well. But for the first time, scientists may have captured snapshots of a false memory in the making.

In computer-enhanced images produced by positron emission tomography (PET), scanning researchers have made pictures of the brain at work recalling a memory and pictures of the brain going awry and bringing up a false memory. The pictures are very similar, in most ways. And to the subjects doing the remembering, both processes feel the same.

But in the images produced by these new brain scans, false memories can be clearly distinguished from those that are true.

Dr. Daniel Schacter of Harvard University led the research team that made the images, which will appear with an accompanying analysis in *Neuron*, a scientific journal. The work is part of a larger effort to decipher the newly urgent phenomenon of false memory.

"We believe we have a clue now about how false memory works," said Dr. Schacter, author of *Searching for Memory* (Basic Books). "This is just a first glimpse of the physical activity underneath the creation of false memories."

Dr. Larry R. Squire, an expert in the neuropsychology of memory at the University of California at San Diego who is not part of the team that produced the new paper, said: "This is important work. It's the first study on the biology of false memory." Combined with another paper that Dr. Schacter has written with more evidence on how false memories work, "this is a landmark in the field of memory," Dr. Squire said.

The work has many implications outside science. The field of memory research was rocked by several criminal prosecutions in the 1980s and early 1990s that were based partly on "recovered memories." One such prosecution was the McMartin preschool case in California, in which Virginia McMartin and six other teachers at her preschool were indicted in 1984 on charges involving ritual sexual and physical abuse of the children, animal sacrifices and satanism. After six years of court fights, some charges were dropped. Mrs. McMartin and her son, Raymond Buckey, were acquitted on some charges, and juries deadlocked on others.

A survey in 1994 by the National Center on Child Abuse and Neglect suggested that more than 12,000 similar accusations—including ritual satanic abuse based on such "recovered memories"—had been brought nationwide, even though, the survey said, none had been substantiated by physical evidence.

Dr. Squire said the McMartin case "was the wake-up call for those of us who study memory." He explained: "Not only lay people, but scientists had to deal with that kind of phenomenon for the first time. We all were presented with this problem: If someone says that kind of abuse happened, we tend to believe them—we think there must be something to it. But there was no other evidence. We all had doubts about whether these memories were real. What could explain it?"

But Dr. Schacter said it was probably far-fetched to think of using PET scans, like those used in his research, in a trial. Dr. Schacter said: "I've had people ask me whether this could be used as a lie detector—you know, just hook them up and see whether their memory is real or not. But I think not. It's all far too complicated, not to mention expensive."

In the study published in the journal *Neuron,* Dr. Schacter, along with Drs. Eric Reiman, Tim Curran, Kathleen McDermott, Henry Roediger and others, used PET scans to observe the brain activity of 12 healthy volunteers as they went through tests in which they formed true and false memories.

PET scans monitor the activation of brain cells by measuring blood flow to the cells. Cells working on making or retrieving a memory use more blood, so those brain areas show up as bright spots on the PET brain image.

In the tests, the volunteers were asked to remember a list of words that were read to them. Then they heard a list of words in which only some of the words had been on the previous list. The volunteers were then asked to identify the words they had heard on the first list; as they tried to remember whether words had been on the list, their acts of recall were recorded with PET scans.

A similar experiment was done for the false memory test, except that when the second list of words was presented orally, some words on the second list were designed to be suspiciously similar, but not the same, as the words studied.

For example, if subjects were presented the words "candy," "cake" and "chocolate" in the first list given in a false memory test, they were later asked if the word "sweet" had been among the ones presented. It had not been, but it was similar enough to prompt many false memories by the subjects.

In many ways, the true and the false recollections showed up in similar ways in the PET scans. Both lit up the area of the brain that is the center of recall: the area adjacent to the left hippocampus, so named by early anatomists after the Greek word for sea horse, which it somewhat resembles.

But something was different about the brain image from the true memory; it had an additional sign of its validity. With the true memory—for example, correctly identifying the word "candy" as one on the first list—subjects had actually heard the experimenter speak the word during the test. So when subjects remembered that word, they had some memory of the way the word sounded.

So PET scans for the true memories showed not only the hippocampal area lit up, but also the left temporal parietal area—near the top of the head and toward the side—where the brain deciphers sound patterns and recognizes words.

A true memory called up both the word and some sensory detail from the moment when the learning had taken place.

In this experiment, it was the sound of the voice uttering the words that was recalled. But Dr. Schacter said that if the test had used lists of words on paper, it might well have been the appearance of the word on paper that was recalled—the shape of the letters, or even the look of the questionnaire or the pencil used to mark the answers.

Many questions remain about how far the technique can be pushed and what its limitations are, Dr. Schacter said. For example, what if the memory of sensory details fades, and rememberers are left with only the gist of what happened, just as with a false memory?

In addition to the first pictures of a false memory, Dr. Schacter and his colleagues have unearthed other features of true and false memories that might be important.

When a person begins a search in memory, Dr. Schacter said, it appears that the effort may bring back more than one item. There are many partial "hits," words that are similar in some way or scenes that have many features in common. Imagined events, as well as memories, can also be netted in the search.

To decide if something is a true memory, the rememberer does some more mental testing to see if something either verifies the memory or excludes it as a match. That extra testing is carried out by yet another area of the brain: the frontal area.

In experiments with four patients trying to remember words after different amounts of studying, Dr. Schacter and his colleagues found that the less a person studied, the more activity was recorded in the frontal lobe of the brain. Conversely, when words had been studied well, it took less searching, so the frontal area was less active.

Older men and women are known to have trouble searching their memories, and they have more false hits in memory tests. When Dr. Schacter and his colleagues carried out an imaging experiment with older people, they found that the trouble with their memories appeared to be not in the hippocampus area itself, the focus of memory, but in the frontal area that is the search engine.

When elderly people were being tested, they showed normal activity in the hippocampal area. But when words were studied only a little, there was little or no additional activity seen in the frontal search area.

One elderly patient with frontal lobe damage was found to be highly susceptible to false recollections. While the hippocampus could generate partial matches and feelings of familiarity with the material, the frontal lobe was apparently unable to do the extra searching that could verify the memory.

In the end, Dr. Schacter said, the new data is a reminder that memory is both complicated and fragile. It is less like looking at videotape and more like making a mosaic from a heap of colored fragments.

—PHILIP J. HILTS, July 1996

Biologists Find Site of Working Memory

WHILE YOU are keeping something in mind, just where exactly is it kept? Neuroscientists, after a long search, think they have an answer. They believe they have located what amounts to the brain's scratch pad, where information is held temporarily when it is needed for some current task.

New techniques for observing the brain in action are revealing that neurons in the prefrontal lobes, just behind the forehead, hold specific kinds of information for short-term use. These neurons appear to be the neural basis for the mind's "working memory" which operates, say, while you dial a phone number or solve a quadratic equation.

The role of the brain's working memory seems similar to that of random access memory (RAM) computer chips, which hold data drawn from the long-term memory systems like a hard drive or a CD-ROM. The cells of the prefrontal cortex can draw data from other regions of the brain, retain the information for as long as needed and switch quickly to other data as the mind's attention shifts elsewhere. "We've found cells for working memory in the prefrontal cortex that retrieve and temporarily hold information pulled from long-term memory stores that are dispersed throughout the brain," said Dr. Patricia Goldman-Rakic, a neuroscientist at Yale Medical School who has done much of the research.

The neurology of working memory, which is more widely known by the now outmoded term "short-term" memory, may hold a vital clue to what goes wrong in the thinking of people with schizophrenia.

"Working memory is the mental glue that links a thought through time from its beginning to its end," said Dr. Goldman-Rakic. "The bizarre thought disorders in schizophrenia, especially the inability to keep a train of thought from getting derailed, could be due to a defect in working memory."

The findings of Dr. Goldman-Rakic and others have shifted brain researchers' attention from the hippocampus, an ancient structure in the limbic system that seems crucial for long-term memory, to the prefrontal lobes, which in evolutionary terms are among the newest parts of the neocortex.

Neuroanatomists have long recognized that the prefrontal cortex is unique in having a huge number of circuits that connect with other parts of the neocortex, especially the centers for analyzing sensory data, and with lower brain centers like the limbic system, which is central to emotional reactions.

The new findings are consistent with theories of neuropsychology, largely based on clinical studies of brain-damaged patients, that view the prefrontal region as the brain's executive center for making decisions, planning and executing behavior. The key executive function of the prefrontal cortex, Dr. Goldman-Rakic argues, is working memory.

While using positron emission tomography (PET) scans, imaging devices that monitor the rate of glucose uptake in the whole brain, she and Dr. Harriet Freedman, also at Yale, trained rhesus monkeys to remember the location of a spot of light that appeared briefly on a television monitor. They found that the task activated a narrow strip of cells in the prefrontal cortex and a zone in the parietal cortex that takes in visual information while the eye tracks an object.

"We've been mapping the activity of neurons throughout the prefrontal cortex," said Dr. Goldman-Rakic. "The common feature of all prefrontal cells we've studied is retrieving information from memory, and each prefrontal area connects to different sensory areas," and so can tap a different kind of memory, such as for bodily sensation or vision.

From another study, based on monitoring single neurons while a monkey retains a visual image in working memory, Dr. Goldman-Rakic is finding that the prefrontal cells are finely tuned for a particular kind of information. Her research team has begun to map the specific neurons involved.

"We've found that the prefrontal zone where cells remember location connects to the area of the visual cortex that specializes in representing spatial relations," said Dr. Goldman-Rakic. "Next to it is another area where cells remember the features of an object, but not its location. This area connects to the temporal cortex, where features are perceived.

"The main centers of the prefrontal cortex are tied to the architecture of the sensory systems," she continued, explaining that this structure allows the prefrontal cortex to retrieve sensory memories.

At Carnegie Mellon University a team led by Dr. Jonathan Cohen is mapping brain activity in subjects who are asked to watch a series of letters flashed on a screen. They respond by pushing a button whenever a letter repeats after a single other letter has intervened—for example, L A L.

Just as the prefrontal cortex has certain areas that specialize in working memory for location and for identity, there are other areas that specialize in working memory for a sequence of objects.

When the letter task was made more complicated, by requiring a response only when two or more letters intervened, progressively larger areas of the prefrontal cortex lighted up in the brain scans.

"The more difficult the task, the more the same areas light up in brain imaging," indicating increased activity, said Dr. Cohen.

Other research teams are taking what amount to moving pictures of working memory as it retrieves and holds data. Dr. James Haxby at the National Institute of Mental Health in Bethesda, Maryland, has used the scanning technique known as functional magnetic resonance imaging to observe people who were asked to look at a picture of a face and hold it in mind for eight seconds, and then look at another picture of a face and compare it with their memory of the first one.

"The visual cortex and prefrontal areas are active when a person first sees the face, but then the visual areas subside while the prefrontal regions sustain their activity during working memory," said Dr. Haxby.

When people were asked to hold the image of a face for up to 21 seconds, an area on the right side of the prefrontal cortex at first was highly active and then faded as time went on, but activity in a region on the left got even stronger over the seconds, Dr. Haxby found.

"We speculate that the right side is involved in holding on to a picture, while the left encodes information about the face—thoughts about it, like who the face might remind you of, or if you find it friendly or sinister," said Dr. Haxby. "As time goes on people's working memory seems to rely more on their analytic understanding of the face than the image itself."

The zones in the prefrontal cortex identified as key to working memory may be but one narrow slice of a much wider array of neurons dispersed throughout the brain, all of which play a role in the process, according to findings by Dr. E. Roy John, director of the Brain Research Laboratory at

New York University Medical Center. Using a non-imaging method to monitor electrical activity of cells throughout the brain, Dr. John believes he can detect more subtle brain activity which, for technical reasons, is missed by imaging methods.

In his research, which monitors electrical activity of cells throughout the brain every eight milliseconds, Dr. John has found patterns indicating that cells from many different areas of the brain are involved in working memory.

"I see working memory widely dispersed throughout the brain," said Dr. John. "There's activity in the prefrontal area and the related parts of the sensory cortex, but also cells in the association cortex, the occipital areas, and other areas. There's a smaller percent of cells in these regions engaged at any given moment compared to those in the prefrontal cortex, but it suggests that working memory is widely distributed."

One clinical implication of the work on the prefrontal cortex and working memory is the potential for a better understanding of some schizophrenia symptoms, such as the derailment of the train of thought.

Blood flow studies of people with schizophrenia have shown that the prefrontal areas are often underactive, and these people also have difficulty in the same visual tracking tasks that Dr. Goldman-Rakic has used in her studies of working memory.

"The prefrontal areas receive a high concentration of dopamine, and drugs effective in treating schizophrenia act on dopamine receptors," said Dr. Goldman-Rakic.

She points out that a relatively new drug for schizophrenia, clozapine, both increases dopamine release in the prefrontal areas and improves patients' thinking and memory.

"The brain's capacity for storing information seems virtually unlimited," said Dr. Goldman-Rakic. "But what good is long-term memory unless we can retrieve it and use it?"

—Daniel Goleman, May 1995

Severe Trauma May Damage the Brain as Well as the Psyche

SEVERE EMOTIONAL TRAUMA may put its victims in double jeopardy, not only searing the psyche, but physically damaging the brain.

New studies in trauma victims as diverse as Vietnam combat veterans and victims of childhood sexual abuse have found a shrinkage in the size of the hippocampus, a brain structure vital to learning and memory. The very hormones that flood the brain to mobilize it in the face of an overwhelming threat can be toxic to cells in the hippocampus, the studies suggest.

"The idea that severe stress or trauma can actually damage the brain is remarkable," said Dr. Dennis Charney, chief of psychiatry at the Veterans Affairs Medical Center in West Haven, Connecticut.

The symptoms of post-traumatic stress include nightmares and vivid flashbacks of the traumatic moment, being fearful and easily startled, having intrusive and disturbing memories of the trauma, irritability and difficulty concentrating.

Combat veterans who still suffer from post-traumatic stress symptoms from the Vietnam War had an 8 percent reduction in the volume of their right hippocampus compared with vets who suffered no such symptoms, according to a report in *The American Journal of Psychiatry* by a team led by Dr. J. Douglas Bremner and including Dr. Charney at the West Haven hospital.

And both combat veterans and and survivors of childhood abuse performed at levels averaging 40 percent lower on a test of verbal memory than did people of comparable age and education, Dr. Bremner said in a report to be published later this year in *Psychiatry Research*.

The traumatized veterans with reduced hippocampus volume were compared with other veterans who had similar backgrounds, body size and other characteristics that might affect brain size, but who had no trauma symptoms.

The left and right hippocampi, twin sea horse–shaped structures deep on each side of the brain, play a crucial role in memory. The hippocampus is especially vital to short-term memory, the holding in mind of a piece of information for a few moments, after which it either resides in more permanent memory or is immediately forgotten. Learning, the accretion of new data in memory, depends on short-term memory.

The finding of shrinkage in the hippocampus suggests a loss of cell mass. Whether the loss is due to the atrophy of dendrites, the long branches that stretch from one brain cell to another, or from the death of such cells, is not yet known.

Some researchers caution that it is not yet certain that trauma and stress shrink the hippocampus. "It might be that people with a smaller hippocampus to begin with are the ones most susceptible to post-traumatic stress symptoms," said Dr. Roger Pitman, a psychiatrist at the Veterans Affairs Medical Center in Manchester, New Hampshire.

"The rule of thumb is that no matter what kind of trauma—an earthquake, combat, rape—only about 15 percent of victims will get long-lasting, severe post-trauma symptoms," said Dr. Bremner.

Dr. Bremner's finding of hippocampus effects is the first to be published, though confirming preliminary results were reported by other research groups at the annual meeting of the American Psychiatric Association in Miami. Dr. Murray Stein, a psychiatrist at the University of California at San Diego, found a 7 percent reduction in hippocampus volume in women who had suffered repeated childhood sexual abuse and who still had post-traumatic symptoms.

Another study of Vietnam veterans found that those who saw more intense combat and suffered from more severe post-trauma symptoms had an average shrinkage of 26 percent in the left hippocampus and 22 percent in the right hippocampus, compared with vets who saw combat but had no symptoms. That finding, also reported at the psychiatry meeting, was by Dr. Tamara Gurvits and Dr. Pitman.

And in new preliminary data, Dr. Bremner also found a loss of hippocampus volume among adults who were victims of repeated physical or sexual abuse as children. Two thirds of the abuse victims were men, and all still had post-traumatic symptoms as a result of the abuse.

"The abuse was extreme," said Dr. Bremner, "including being raped or otherwise sexually penetrated regularly, or being routinely hit, burned or otherwise physically abused over a year or longer."

The memory of those with post-traumatic symptoms is particularly faulty for words, like grocery lists or phone numbers. Using a test in which people hear a story and then repeat it back immediately and again 15 minutes later, Dr. Bremner found a 40 percent deficit in the accuracy of short-term memory in trauma victims compared with people of similar age and background.

The study found no deficiency in the trauma victims' overall IQ scores, nor in other kinds of memory, like that for things they had seen or for vocabulary.

"The hippocampus is crucial for short-term verbal memory," said Dr. Bremner.

The shrinkage in the hippocampus may be due to the effects of heightened levels of cortisol, a steroid hormone secreted by the brain in response to emergencies, some researchers believe. "Cortisol can be toxic to the hippocampus," said Dr. Bremner.

Studies in rats and primates suggest that glucocorticoids, the equivalent of cortisol in other species, "may be neurotoxic to the hippocampus at the massive levels that are released under extreme stress or during trauma," said Dr. Robert Sapolsky, a neuroscientist at Stanford University. "I'm talking about the levels you would see in a zebra running from a lion, or a person fleeing a mugger—a real physical life-and-death crisis—if it is repeated again and again as time goes on."

During such moments of crisis, the body engages in a kind of biological triage, suppressing less pressing functions like ovulation, immune-system reactions and growth, shunting its resources to the muscles needed to survive the immediate danger.

Cortisol is a major means the body uses, with adrenaline, to arouse itself so quickly; its action, for example, triggers an increase in blood pressure and mobilizes energy from fat tissue and the liver.

"The dark side of this picture is the neurological effects," said Dr. Sapolsky. "It's necessary for survival, but it can be disastrous if you secrete cortisol for months or years on end. We've known it could lead to stress-exacerbated diseases like hypertension or adult-onset diabetes. But now we're finding the hippocampus is also damaged by these secretions."

Studies in animals show that when glucocorticoids are secreted at high levels for several hours or days, there is a detectable effect on memory, though no neuronal death. But with sustained release from repeated stress, "it eventually kills neurons in the hippocampus," said Dr. Sapolsky. "This has been shown solidly in rats, with the cell biology well understood."

A parallel effect has long been known among patients with Cushing's disease, a hormonal condition in which tumors in the adrenal or pituitary glands or corticosteroid drugs used for a prolonged time cause the adrenal glands to secrete high levels of a hormone called ACTH and of cortisol. Such patients are prone to a range of diseases "in any organ with stress sensitivity," including diabetes, hypertension and suppression of the immune system, said Dr. Sapolsky.

Cushing's patients also have pronounced memory problems, especially for facts like where a car was parked. "The hippocampus is essential for transferring such facts from short-term to long-term memory," said Dr. Sapolsky.

In 1993, researchers at the University of Michigan reported that magnetic resonance imaging (MRI) had shown an atrophy and shrinkage of the hippocampus in patients with Cushing's disease; the higher their levels of cortisol, the more shrinkage.

In an apparent paradox, low levels of cortisol in post-trauma victims were found in a separate research report, also in *The American Journal of Psychiatry.* Dr. Rachel Yehuda, a psychologist at Mount Sinai Medical School in New York City, found the lower levels of cortisol in Holocaust survivors who had been in concentration camps 50 years ago and who still had post-traumatic symptoms.

"There are mixed findings on cortisol levels in trauma victims, with some researchers finding very high levels and others finding very low levels," said Dr. Sapolsky. "Biologically speaking, there may be different kinds of post-traumatic stress."

In a series of studies, Dr. Yehuda has found that those post-trauma patients who have low cortisol levels also seem to have "a hypersensitivity in cell receptors for cortisol," she said. To protect itself, the body seems to reset its cortisol levels at a lower point.

The low cortisol levels "seem paradoxical, but both too much and too little can be bad," said Dr. Yehuda. "There are different kinds of cells in various regions of the hippocampus that react to cortisol. Some atrophy or die if there is too little cortisol, some if there is too much."

Dr. Yehuda added, "In a brain scan, there's no way to know exactly which cells have died."

To be sure that the shrinkage found in the hippocampus of trauma victims is indeed because of the events they suffered through, researchers are now turning to prospective studies, where before-and-after brain images can be made of people who have not yet undergone trauma, but are at high risk, or who have undergone it so recently that cell death has not had time to occur.

Dr. Charney, for example, is planning to take MRI scans of the brains of emergency workers like police officers and firefighters and hopes to do the same with young inner-city children, who are at very high risk of being traumatized over the course of childhood and adolescence. Dr. Pitman, with Dr. Yehuda, plans a similar study of trauma victims in Israel as they are being treated in emergency rooms.

Dr. Yehuda held out some hope for people who have suffered through traumatic events. "It's not necessarily the case that if you've been traumatized your hippocampus is smaller," she said. She cited research with rats by Dr. Bruce McEwen, a neuroscientist at Rockefeller University, showing that atrophied dendritic extensions to other cells in the hippocampus grew back when the rats were given drugs that blocked stress hormones.

Dr. Sapolsky cited similar results in patients with Cushing's disease whose cortisol levels returned to normal after tumors were removed. "If the loss of hippocampal volume in trauma victims is due to the atrophy of dendrites rather than to cell death, then it is potentially reversible, or may be so one day," he said.

—DANIEL GOLEMAN, August 1995

Doctors Record Signals of Brain Cells Linked to Memory

A TEAM OF NEUROSURGEONS say that in preparing epilepsy patients for operation, they have recorded signals from single human brain cells that make the associations leading to long-term memories, the first time that feat has been achieved.

In some cases the cells seemed to know better than their owner, since they responded positively to a photograph of a face that the researchers had already shown but that the patient denied having seen before.

The tests were performed by Drs. Itzhak Fried, Katherine A. MacDonald and Charles L. Wilson of the University of California at Los Angeles in the course of pinpointing the source regions of epilepsy in a series of patients. The results were published in the journal *Neuron*.

The nerve cells were situated in a structure of the brain called the hippocampus. Patients whose hippocampus has been removed retain memories from before the operation but cannot form new ones, suggesting that the hippocampus is required for forming long-term memories but is not a site of permanent storage.

Ethical opportunities to record from the human hippocampus are very rare, and a previous effort did not capture the associations that, given past work with rodents and monkeys, Dr. Fried and his colleagues thought they would see.

New methods of scanning the brain have helped locate many of its functions, but these scanning methods measure blood flow to active regions and do not reveal just how brain cells encode information. That requires recording directly from a nerve cell by sticking an electrode into it, as the Fried team did.

An emerging feature of knowledge about the brain is that incoming information is separated into many different strands. For instance, faces, letters and colors are processed in different areas of the cortex, the thin sheet of nerve cells that makes up the outer surface of the brain. Even for faces, special attributes like identity, expression and sex are represented in different parts of the cortex. An outstanding problem for brain scientists is to explain how these separately processed attributes are brought together.

In terms of long-term memory, the hippocampus, a structure with intimate connections to many regions of the cortex, has long seemed cast for a leading role in bringing together separate attributes.

Studies with rats and mice have uncovered the remarkable phenomenon of what are called "place cells" in the hippocampus. When the mouse enters a strange environment, many hippocampal cells are recruited to record the new scene. Researchers recording signals from the cells can tell where the mouse is in its cage by looking to see which cell is firing.

Discovery of the place cells sets up a debate among neuroscientists. Some say that forming spatial memories is the main role of the hippocampus. Others contend that this may be true in rats and mice, for which space is important, but that in humans the hippocampus has a much broader role.

The new data reported by Dr. Fried and his colleagues support the second view. Figuring that humans are very oriented toward recognizing individual faces and their expressions, and that faces would be as important to humans as spaces are to rats, Dr. Fried showed a set of faces expressing different emotions to his wired-up patients. He recorded responses from the hippocampus cells both when the faces were being identified for the first time, and when they were recognized or otherwise on a second showing.

A fair number of the cells responded singly to conjunctions of various attributes. For example, one cell responded to the expression of anger in two different faces. Others responded strongly to faces that were both new and wearing angry expressions.

"This is the first time in the human that we have seen this type of conjunctive coding," Dr. Fried said.

Of particular note were cells that responded positively to a previously seen face, even though the patients said they did not remember having seen the face before. In possible explanation of that strange result, Dr. Fried said that "we are essentially probing into very large networks of cells" and that the conscious decision of whether a face has been seen before "is probably a matter of voting" among the cells. In other words, the cell this electrode had happened to hit said aye on the question of recognition but may have been out-voted by other cells in the network.

Several experts said the new result was important because it established that the hippocampus has a role in humans similar to that found in rodents and monkeys, thus confirming that these animals are good guides for studying the human brain.

"It's a very nice confirmation of the similarity across species," said Dr. Howard B. Eichenbaum of Boston University.

—NICHOLAS WADE, May 1997

Very Smart Fruit Flies Yield Clues to the Molecular Basis of Memory

IN BOTTLES LINING A WALL of a Long Island laboratory there are swarms of fruit flies with an unusual ability. They have been endowed with a gene that gives them photographic memory.

In bottles nearby are their less fortunate cousins, genetically engineered for forgetfulness. And one floor below scamper another product of the genetic engineer's art: amnesiac mice.

These flies and mice are the product of efforts to identify the genes and molecules that are involved in laying down long-term memory. Researchers have found a protein that serves as a kind of logical switch, signaling to the nerve cell whether a memory is to be stored for a fleeting moment or permanently engraved in the mental archives.

This protein switch has its counterparts in flies, mice and humans. "At a nuts-and-bolts level, our brains are working by the same principles and mechanisms as those of little fruit flies," said Dr. Alcino Silva, a neuroscientist at Cold Spring Harbor Laboratory on Long Island who has led much of the mouse work.

Indeed, the recent work on this switch, called CREB, has given scientists "a new vantage point for understanding how memory works," said Dr. Eric Kandel, a neurobiologist at Columbia University's College of Physicians and Surgeons in New York City who has pioneered research on the molecular basis of memory. Many molecules, he noted, are involved in governing something as complicated as long-term memory. But CREB has afforded the most enticing clue to the mystery of how the brain decides what it will and will not remember for good.

"CREB is the clearest example of a molecule involved in long-term memory" to come out of behavioral studies, said Dr. Larry R. Squire, a neuroscientist at the Veterans Affairs Medical Center in San Diego.

Dr. Howard B. Eichenbaum, a neuroscientist at Boston University, said: "I'm very excited. It's amazing that CREB is so specific to memory.

"The CREB story is growing stronger as new evidence" provides powerful links between the protein and various memory processes, he said.

The discovery of CREB's role in fruit flies and mice has far-reaching implications. It could answer such questions as why cramming for a test does not work in the long run or why certain emotional events become instantly etched in the mind. Medically, the findings could possibly lead to drug treatments for memory loss, dementia and post-traumatic stress disorder.

When the CREB switch in a cell is turned on, researchers believe, it sets off the synthesis of other proteins that cement lasting memories by supporting the growth of new connections between nerve cells. When it is turned off, CREB halts the production of those cementing proteins, thus preventing unnecessary memories from forming.

Studies done in Dr. Kandel's laboratory on sea slug cells supplied the first hint of a role for CREB in memory. But the recent fruit fly work provides the most striking behavioral demonstration that CREB works as a memory switch.

In fruit flies, as in other species, CREB is a so-called transcription factor, a protein in the cell nucleus that binds to DNA and causes nearby genes to be spun into protein. Researchers have discovered how the nerve cell flips the CREB switch on and off. A protein called the CREB activator turns it on, and CREB repressor turns it off.

The gene sequences used to make the CREB activator and CREB repressor proteins have also been identified, and a few years ago Dr. Jerry Yin, a biologist now at Cold Spring Harbor, endowed fruit flies with extra genes so that one group acquired an extra CREB activator and the other gained a CREB repressor. To test their memory, he teamed up with Dr. Timothy Tully, a geneticist at Cold Spring Harbor.

Dr. Tully developed a test that measures how fast the flies learn to associate an odor with an electric shock in a way that produced a lasting memory. Normal flies need 10 training sessions to form a persistent recollection of the test. Flies with an extra dose of CREB repressor could not form lasting memories at all. "That showed beyond reasonable doubt" that CREB repressor blocks long-term memory, Dr. Tully said.

But most surprising of all, the insects fortified with an extra CREB activator gene needed just a single training session. "This implies these flies have a photographic memory," Dr. Tully said. He said they are just like students "who could read a chapter of a book once, see it in their mind and tell you that the answer is in paragraph 3 of page 274."

The state of the CREB switch, at least in fruit flies, seems to depend on the prevailing balance in the nerve cell between supplies of CREB activator and CREB repressor. A preponderance of CREB activator is needed for memory storage, said Dr. Tully, who published his and Dr. Yin's results last year in the journal *Cell*.

Ordinarily, there is an equilibrium between activator and repressor, researchers believe. CREB repressor remains present, they suspect, to prevent the storage of boring and unnecessary detail—the clutter in a room, the babble in a bar, the "ums" in a spoken sentence. "Memory is not about storing information; it's about storing useful information," Dr. Silva explained.

The CREB repressor can be thought of as a memory filter. It dominates, the theory goes, until something important happens, like an emotionally powerful event, that either removes CREB repressor from nerve cells or increases the levels of CREB activator enough to make brain cells lay down a permanent memory. This is presumably the mechanism by which people vividly remember where they were when President John F. Kennedy was assassinated or, as in Dr. Silva's case, seeing a little red bicycle he wanted at the age of 5.

Dr. Silva has recently moved the fruit fly work forward by studying a similar system in experimental mice. Mice learn what is safe to eat by smelling what is on one another's breath, behavior that Dr. Silva exploited to measure his mice's ability to remember what they learn. He has found that mice with a defect in the CREB activator gene that causes them to make much less of its product than is normal are virtually unable to form long-term memories. His article is to appear in the journal *Current Biology*.

Dr. Silva also discovered that his forgetful mice could be made to remember much better when they had short lessons with rests in between. The treatment looks a lot like what good students do—study in many short bouts instead of cramming just before a final. In both cases, Dr. Silva suggested, the small amount of available CREB activator in the relevant brain cells may limit the amount of information an animal, or a person, can take in at one time.

Shorter bouts of learning separated by rest, he proposed, allow time for the available activator to recycle from the previous learning trial and respond again—a molecular argument for steady studying. "We can now give you a biological reason why cramming doesn't work," Dr. Tully said.

He and others also hope to find chemical ways of enhancing brain cell function in people with dementias like Alzheimer's disease and even age-related memory loss. Dr. Tully and Dr. Yin are forming a company called Helicon Therapeutics to parlay their knowledge of CREB into pharmaceutical products.

Of course, such products must depend on knowledge of many molecules other than CREB. "It's hard to link such a complicated process as learning and memory to just one molecule," said Dr. Richard Goodman, a neuroscientist at Oregon Health Sciences University in Portland.

Others agree and are seeking to identify the molecular machinery surrounding CREB, including the thousands of proteins with genes that CREB controls. They are also trying to link molecular memory processes to larger scale changes in brain cells and brain cell circuits.

Indeed, many secrets of memory seem poised to unravel from work on CREB. "CREB is one of the first truly solid molecular clues about memory," Dr. Silva said. And memory, Dr. Kandel added, is "who we are."

—INGRID WICKELGREN, December 1996

Two Studies Suggest Sleep Is Vital in Consolidating Memories

IN SEPARATE EXPERIMENTS conducted on rats and humans, scientists have found the first direct evidence that memories are consolidated during sleep.

The findings, which are being reported in the journal *Science,* have implications for students cramming to pass an examination, people learning new skills like skating and anyone trying to navigate through new surroundings. They suggest for example, that people learning to ski might improve their performance after a few nights' sleep.

In the rat studies, researchers found that when an animal moved through a somewhat familiar environment, a particular pattern of cells was activated in the hippocampus. Soon after, when the rat was allowed to sleep, the same cells showed an increased tendency to fire at the same time. The researchers hypothesize that this occurs as the memories are being consolidated in the brain.

The human studies, conducted by another group of researchers, focused on a different type of memory, involving the learning of a new skill. When the subjects were allowed to sleep and then repeatedly roused to wakefulness when they entered a period of vivid dreaming called rapid-eye-movement sleep, it hampered their performance on subsequent tests of the skill. This provided the best evidence so far that memories are consolidated during sleep.

The animal experiments were conducted at the University of Arizona in Tucson by Dr. Matthew Wilson and Dr. James McNaughton. The human experiments were carried out by Dr. Avi Karni and Dr. Dov Sagi at the Weizmann Institute in Rehovot, Israel.

The Arizona group was interested in spatial memories. Animals have to know where they are at all times so they can hunt, explore new territories and find their way home, said Dr. Wilson, who has just moved to the Massachusetts Institute of Technology. In animals and humans, he said in a telephone interview, spatial memories are first formed in the hippocampus. Humans also use the hippocampus to acquire and store so-called declarative memories, involving all the events with which a person is involved during waking hours.

Most scientists believe that the hippocampus is essential for processing and consolidating new memories, Dr. Wilson said. They suspect that hippocampal neurons somehow strengthen the links between neurons in other regions of the brain where various aspects of each memory are stored; when a memory is recalled, these links are activated simultaneously and the memory is perceived as a coherent whole.

According to this hypothesis, specific cells in the hippocampus would be primed by events to help encode a new memory. But just how the information held temporarily in the hippocampus is shuffled out to other brain regions for long-term storage is not known.

In a series of experiments, Dr. Wilson implanted electrodes into so-called place cells in the rat hippocampus. Each place cell is activated when an animal is in a certain region in its environment, Dr. Wilson said. A place cell that fires when an animal is in one position will stop firing when the animal moves to a different place. As an animal moves through an environment, like an enclosed box with distinct features on the walls, different ensembles of place cells will fire to reflect the animal's movements through that space, he said.

Rats fitted with a lightweight multi-electrode array were placed in an enclosed box that they had been in before but that they did not know well. They were then allowed to explore, as the electrodes recorded the activity of their place cells. Next the rats were put back in their cage, where they fell back asleep. After five to ten minutes, Dr. Wilson said, the rats entered slow-wave sleep, which is dream-like but does not contain vivid dreams.

Hippocampal place cells that fired together when the rats explored the box showed an increased tendency to fire together during sleep, Dr. Wilson said. Information about activity in the environment is preserved in the hippocampus with the same ensembles of cells, he said, and it is acted upon by the sleeping brain.

Exactly what happens next is not known, Dr. Wilson said. The leading hypothesis is that the hippocampus plays out the stored information to memory networks throughout the brain. During this reorganization and consolidation of memories, he said, the brain creates a more useful representation of information than what was stored during the actual experiences.

Humans also have place cells, Dr. Wilson said. When people learn to navigate through a new environment, they presumably form patterns in their hippocampi that are later processed in sleep and stored throughout the brain. Thus ability to find one's way around a new city should improve after a night's sleep.

The same may be true for students who cram for an examination. Experiments done in Canada show that students who get some sleep after studying for an exam retain more information than those who stay awake overnight. While short-term memory can hold a lot of information, it may be that sleep helps consolidate memories in ways that are more useful to the student.

Dr. Karni, who helped carry out the human experiments, explored a different phenomenon called procedural memory, which involves repetitive movements that help people learn to perform skills, like riding a bicycle, skating or downhill skiing.

Because procedural memories are not processed through the hippocampus, most researchers felt that sleep would have no effect on them. Dr. Karni's experiments tested this notion for the first time.

In the experiments, subjects were trained in a simple visual task: to recognize the orientation of an object appearing in their peripheral vision. After several days' practice, they got progressively better, Dr. Karni said. Last year he reported that once people learned this skill, they retained it for years.

But while doing those experiments, Dr. Karni, who is now at the National Institute of Mental Health in Bethesda, Maryland, noticed that subjects improved after an eight-hour delay. To see if sleep played a role in their better memories, subjects were trained in the evening before they went to sleep. One group was awakened every time they began to enter rapid-eye-movement sleep. They were roused about 60 times during the night. Members of a second group were awakened every time they entered slow-wave sleep, a stage when dreams tend not to occur. They were roused a comparable number of times during the night.

With rapid-eye-movement or dream-sleep deprivation, the subjects' learning did not improve at all, Dr. Karni said. They did as well as they had the night before. But subjects who were awakened during periods of slow-wave sleep, but who were allowed to dream, showed distinct improvement over the night, he said.

Thus dreaming seems important for consolidating procedural memories, Dr. Karni said. Exactly what happens during dreaming to improve performance on procedural memory tasks is not known, he said.

—SANDRA BLAKESLEE, July 1994

4

LANGUAGE AND THE BRAIN

Language is the most distinctive power of the human brain. Other animals communicate by chemical or visual signals, but only humans can string together long series of signals under the constraint of grammatical rules that make one individual's utterances comprehensible to another.

Several psychologists have made strenuous efforts to teach language to chimpanzees. The experiments produced fascinating instances of chimp-human communication but no evidence of true language. The chimps, wonderful mimics, became adept at behaviors that induced their human handlers to give over food. But the behaviors fell far short of even a young child's ability to combine symbols in meaningful sentences.

The language ability seems therefore to be been acquired quite recently in human evolution, and certainly after the divergence of the hominid and chimpanzee lines some 5 million years ago.

It may be that some components of language, such as reading and writing, developed even more recently than speech and hearing. In any event, all these aspects of language seem to be mediated by different, specialized regions of the brain.

Traffic Jams in Brain Networks May Result in Verbal Stumbles

IT IS SO FRUSTRATING! You are talking to a friend when suddenly you cannot remember the name of something. It is on the tip of your tongue but no matter how hard you try, you cannot say what it is. Even more maddening, you know so much about it: It is an animal that lives in South America. It gives wool that is sometimes made into sweaters. You met someone in California last year who raises them. It is an . . . er, um . . .

To scientists who study the brain, this is a tip-of-the-tongue experience. It even crops up in a slightly different form among users of sign language. They call it a tip-of-the-finger experience.

"Humans love to talk," said Dr. Willem Levelt, director of the Max Planck Institute for Psycholinguistics in Nijmegen, the Netherlands. "Most of us spend large parts of the day in conversation. If we are not talking to others, we are talking to ourselves."

Dr. Levelt's interest is in the problem of how people go from thinking about something to actually saying it. In the course of studying the component systems involved in generating spoken words from thought, he has developed a theory about what is happening inside the brain when a speaker blocks on a word.

The identification of these systems is based on advanced techniques for imaging the brain and on precise timing of the brain's activities, he said. These two techniques allow researchers to watch the process whereby thoughts are transformed into a seamless flow of words, to see where this process breaks down and to show how people correct speech errors on the fly.

Dr. Levelt and his colleagues base their research on the assumption that the human brain contains distinct modules for processing thought into language. Dr. Gary Dell, a professor of psychology at the University of Illinois at Champaign-Urbana, said that this assumption is widely accepted among psycholinguists, though it is expressed in slightly different forms.

The exact anatomy of the proposed modules is not yet known, Dr. Levelt said in a telephone interview, but their existence can be demonstrated experimentally. They are not little boxes in the brain, he said, but widespread networks of interconnected neurons that cooperate, with precise timing, to carry out specific tasks.

Dr. Levelt calls the three modules the lexical network, the lemma network and the lexeme network. Essentially, the lexical network handles thoughts, the lemma network handles syntax and the lexeme network manages spoken sounds.

In speaking, the first module to be activated is the lexical network. "Conceptualizing is deciding what to express, given our intentions," Dr. Levelt said. "As speakers, we spend most of our attention on these matters of content" and how to order thoughts sequentially.

Once a message is thought out, he said, "we must capture it by some lexical concept." To do so, we dip into our stored vocabulary—typically tens of thousands of words encoded by neurons. Speakers can retrieve two to three words per second containing 10 to 15 sounds or syllables.

Individual words and concepts are stored in vast interactive networks, called nodes, throughout the brain, Dr. Levelt said. Moreover, they are distributed differently in each person using multiple regions of the brain. Language networks are as distinct as fingerprints; no two people are identical.

Imagine you want to say the word "llama," Dr. Levelt said. Perhaps you saw a picture of a llama, or you thought of the animal while talking with a friend. The mind first activates the lexical node for llama, which contains everything you know about llamas. It is an ungulate with a long neck, it is used as a pack animal and so forth.

When the lexical node for llama is activated, nodes for words of similar meaning are also stimulated, Dr. Levelt said. These might include the nodes of sheep and goats, nodes for beasts of burden in general, the node for hoofed animals and so forth.

At this point, you still do not have the word for llama, Dr. Levelt said. But you have activated a great deal of information about llamas and similar animals.

The next stage in processing is handled by the second module, the lemma network. When the lexical concept for llama and other activated concepts are passed to this level, two things happen.

First, the lemma assigns proper syntax to each incoming concept. These are the rules of the speaker's language, including word order, gender if appropriate, case markings and other grammatical features. Also at the lemma level, verbs, nouns and modifiers are put in their proper places in a word string.

Second, the various activated lexical concepts engage in a competition. Most of the time, the most highly activated concept (llama) will win. But sometimes there is interference from other lexical concepts. The more that are activated, the longer it takes to generate the desired word, Dr. Levelt said.

Timing experiments done in Dr. Levelt's laboratory show how this works. Subjects are shown pictures and asked to name them as fast as they can. The average naming time is 70 thousandths of a second. Then the experimenter adds a distraction—such as muttering the word horse when a picture of a cow is shown. People need 80 thousandths of a second on average to name cow when the lexical concept of a horse is also activated, Dr. Levelt said. But if an unrelated lexical concept like house is muttered, there is no decrease in the time in takes to say cow.

The third part of the process is to turn a chosen lexical concept into a spoken word.

This is called the lexeme level, Dr. Levelt said. "We store thousands of phonemes, which are the highly practiced sounds of a language—the 'la' and 'ma' sounds that make the word llama. The sound pattern of a language—its tones, metrics and intonation—is also stored at the lexeme level.

"Accessing the lexeme level is harder than you think," Dr. Levelt said. The mind has to find the correct sounds and match them to the syntactical elements processed by the lemma network.

This is where the process of generating thoughts into speech can fail, Dr. Levelt said. Many things can go wrong.

One is the tip-of-the-tongue phenomenon. You are at the lemma level but the word refuses to come, Dr. Levelt said. You know a lot about it. You might even know it has two syllables with stress on the second syllable, which suggests that part of the lexeme information is accessible.

People who speak languages with masculine and feminine words will almost always know the correct gender of the missing word, said Dr. William Badecker, a research scientist at Johns Hopkins University who works with brain-damaged patients who have trouble naming things.

Deaf people who use sign language experience the same frustration, said Dr. Ursula Bellugi, a cognitive scientist at the Salk Institute in San Diego. They know a lot about the missing word but cannot make the proper hand shape or may have trouble deciding which direction the hand should move, she said.

Why do words become blocked?

"We don't know for certain," Dr. Levelt said. But one idea is that a given lexical node is not sufficiently activated to spread to the lexeme level where syllables are stored. Thus llama might not win out over goats and sheep, leaving the speaker fumbling for the word.

Experiments suggest that the familiarity of a word plays a role in tip-of-the-tongue experiences, Dr. Levelt said. Words used less often take longer to name. For example, it takes 200 thousandths of a second longer to say the word moth when shown a picture of it than to say mouth.

Many people block on names, especially ones that do not crop up frequently in daily conversation, Dr. Levelt said. It seems that the more unique a name is, the harder it is to recall. But a name like Baker, which may be stored in a wider network, is easier to recall.

People also complain that their memories for names and words decline with age. "We don't think transmission speed is slower in older brains but the number of active connections between cells may be less," Dr. Levelt said. The activation process may spread more slowly.

But as most people have discovered, waiting a few minutes (or longer) will help retrieve a word on the tip of your tongue.

Dr. Dell explained what happens: "Say you are trying to remember the name of that funny stuff inside a sperm whale, the stuff used in perfume. You may think it sounds like amber but you know it's not. But you keep thinking, amber, amber, which makes the amber part of your network activate. Eventually, if you are not successful at finding the wanted word, you give up, think about something else, throw in a little randomness. Later, when you think again, the word may suddenly appear—ambergris." Thus waiting gives the brain the time to "re-boot" its lexical concepts so that the correct one can try again to win the competition, he said.

Other kinds of speech errors can occur in the transition of a thought between the lemma network and the lexeme network, Dr. Dell said. Sometimes people exchange one word for another (*fill up my gas with car*) or mix up speech sounds (*queer old dean* instead of *dear old queen*).

So-called Freudian slips of the tongue are also common. Freud thought that they represent deep sexual urges but they are more innocent, Dr. Levelt said. While talking, people are often thinking about other things, which can cause an unrelated lexical node to become activated. Thus a word from a different semantic area can drop into the conversation, surprising both the speaker and listener.

But if all goes well and a word is retrieved correctly, it goes to next level of processing which is articulation, Dr. Levelt said. This is the process whereby the syllables are mapped onto motor patterns generated in the tongue, mouth, lips, larynx and lungs.

It is at this point that people can correct errors in their speech, Dr. Levelt said. "Sometimes we edit our speech on the fly," Dr. Levelt said, using "um" and "er" to signal that trouble is afoot. "And when we restart the sentence, we do so in such a way that the proper syntax is preserved," he said.

Thus a person might say, "Is the nurse, er, the doctor interviewing patients?" but is less likely to say "Is the doctor interviewing, er, the doctor seeing patients?" The second correction is clumsier and people tend to repair errors in the right word order, Dr. Levelt said. This is why it is usually possible to splice out speech errors in the broadcast business.

The thought-to-talk process has been studied most in terms of its timing. But where exactly is it occurring in the brain? Most of the conceptual networks are widely distributed, Dr. Levelt said, but an experiment to be conducted next month in Helsinki may shed light on where the lemma-to-lexeme process occurs.

Using a brain imaging machine called a magnetoencephalograph, subjects will be shown pictures with low- and high-frequency names. For example, it takes longer to say broom than it does to say boat. By comparing where in time and space the two words deviate and overlap, researchers hope to find where meaning is turned into words.

The brain evolved two separate systems that were eventually stitched together to facilitate spoken language, Dr. Levelt said. One level involves cognitive acts such as thoughts and syntax; these are the conceptual and lemma modules. The second level involves motor acts such as pronouncing syllables and stringing sounds into words; these are the lexeme and articulation networks. When babies learn to talk, they develop the two systems more or less independently until age 2, when the systems merge. Most speech breakdowns, including tip-of-the-tongue experiences, occur at the interface between these two systems, Dr. Levelt said.

—SANDRA BLAKESLEE, September 1995

Workings of Split Brain Challenge Notions of How Language Evolved

LAST YEAR, a 43-year-old woman who had suffered from terrible seizures since she was 16 underwent surgery to cut the thick band of fibers connecting the left and right hemispheres of her brain. The treatment worked as hoped, stopping epileptic seizures from spreading throughout her brain and protecting her from the head injuries and burns she had suffered from falls when she suddenly lost all muscle control.

There was only one problem. After the surgery, the woman, identified in research reports only as V. J., could no longer write—not her name, not familiar phrases, not at all. But she could speak and understand spoken language with no difficulty.

Splitting the two halves of the brain surgically has given neuroscientists many insights into the workings of the two hemispheres. This particular accident of surgery, says a prominent neuroscientist, challenges standard theories about how language abilities—speaking, reading and writing—evolved and are organized in the brain.

What is significant in V. J.'s case is that her writing activity seems to be controlled completely by the right brain hemisphere. The prevailing theory, supported by earlier findings in patients with the two hemispheres separated, is that with very few exceptions virtually all language abilities are found in the left hemisphere, said Dr. Michael Gazzaniga, head of the cognitive neuroscience program at Dartmouth College in Hanover, New Hampshire.

This is true even for people who are left-handed, like V. J., even though the muscular activity of writing with the left hand is controlled by the right brain hemisphere. This is the first time that speaking has been shown to lie on one side of the brain and writing on the other, Dr. Gazzaniga said. V. J. apparently has some reading ability on both sides of the brain.

This dissociation, described at a Society for Neuroscience meeting in Washington, has thus far been shown only in this one person, but may well apply more broadly to other left-handed people, Dr. Gazzaniga said. In the brain sciences, he said, insights from one case have often led to general principles.

The study of V. J. was carried out by Dr. Gazzaniga and a colleague, Dr. Kathleen Baynes, an assistant professor of neurology at the University of California at Davis Center for Neuroscience.

That talking is in the left and writing in the right side of the brain in this woman is "really fascinating," said Dr. Steven Pinker, a linguist at the Massachusetts Institute of Technology. It suggests that reading and writing arose separately from spoken language and may be wired up in the brain wherever there are "spare areas," he said. In this case, the hookups for writing were made in the right hemisphere.

"It's a very interesting observation," said Dr. Marcus Raichle, an expert on brain imaging at Washington University in St. Louis. There has never been a systematic effort to study left-handed people using brain imaging techniques, he said, and so no one has really tried to ferret out the circuitry for writing as opposed to reading and speaking.

The standard theory has been that language evolved about 100,000 years ago, and the improved communication helped those individuals who had it thrive. The ability to speak thus spread quickly through the human species. Reading and writing arose less than 10,000 years ago, presumably as a cultural invention taking advantage of an underlying biological ability that was already in place, not as part of biological evolution.

The predominant idea has been that reading and writing were laid on top of speech in the left hemisphere, tightly interconnected. But the new research suggests that these human inventions can hop around in the brain, even to the opposite side, Dr. Gazzaniga said.

In all people, the right side of the brain controls the left half of the body. The left side of the brain handles the right side of the body. Similarly, in vision, the right hemisphere receives information from the eyes about all objects and movements detected in the left visual field, the space to the left side of the head. Information from the right visual field is transmitted to the left hemisphere. Both halves share and integrate information to give a complete picture of the world.

In most humans, including congenitally deaf people who use sign language, all language abilities—speaking, reading and writing—are grounded in the left hemisphere.

About 10 percent of people are left-handed, Dr. Gazzaniga said. Eighty percent of them have all language abilities localized in the left hemisphere, just like righties. Studies have shown that the motor ability of the left hand (and right hemisphere) for writing is still dependent on the left hemisphere's language ability.

The other lefties fall into two groups, Dr. Gazzaniga said. Some are completely flipped in terms of hemispheric dominance—all language functions are in the right side and spatial abilities are in the left side. And the last group has shown a puzzling mixed pattern, perhaps like that of V. J. But nobody has tracked these exceptions carefully enough to analyze where, exactly, language abilities lay, until this case, Dr. Gazzaniga said.

V. J. is the first person to shed real light on the problem, said Dr. Baynes, Dr. Gazzaniga's colleague. Dr. Baynes put V. J. in front of a screen onto which words—nouns, verbs and adjectives—were flashed to each side of her brain independently. Her hands rested below the screen, shielded so that her eyes could not see them. Each hand held a pen over a tablet of paper.

"When a word appeared, V. J. was instructed to write what she saw on the screen and to tell us what she saw," Dr. Baynes said.

When a word was shown to the left hemisphere—the one with spoken language—she could read the word, say it and spell it out loud, Dr. Baynes said. But she could not write it down. Attempts at writing were illegible.

(Because her brain was disconnected, she would have to make an effort to use her right hand to try to write what she saw in this part of the experiment, Dr. Baynes said. It was difficult to imagine this, but her left side and left hand were totally unaware of what was going on.)

When words were shown to the right hemisphere, V. J. was stumped, Dr. Baynes said. She would look at the word and say, "Um, I think there's something there but I can't tell what it is." She could not read, speak or spell the words. But amazingly she could write them down.

"She'd pick up the pencil and boom, write out the words, no problem," Dr. Gazzaniga said. "It is just astounding. Here is the executive writing system acting outside the system that can actually speak with all the usual phonological mechanisms.

"At what level is the right hemisphere able to do this?" Dr. Gazzaniga continued. "It may be simple copying and not a real expression of normal writing." But in further tests, not yet submitted for publication, it appears that the true writing system is involved, he said.

The next step will be to test V. J. in more detail, Dr. Gazzaniga said, "to see what she can and cannot do." It would also be nice to begin testing left-handed people with functional magnetic resonance imaging, which detects individual differences in brain wiring, he said. V. J. may be a very rare case or she may be proof of a larger phenomenon—that newer linguistic skills like writing will be laid down wherever the brain has spare room.

—SANDRA BLAKESLEE, November 1996

"Glasses for the Ears" Easing Children's Language Woes

SCIENTISTS HAVE DEVELOPED a radically different treatment for children with severe language and reading difficulties, one that may have applications for millions of children with dyslexia. They call it "glasses for the ears."

The treatment uses a special form of computer-generated speech to train the children to hear differences in sounds that they could not hear before. The researchers believe their program actually results in changes in the parts of the brain that process simple sounds. Unlike eyeglasses, the treatment is designed to produce permanent changes in the ability to understand spoken and written language.

Recent experiments show that after just four weeks of treatment, language-disabled children advanced two full years in their verbal comprehension skills, researchers say. They said the improvements endured after training had stopped. In effect, the children could throw their "glasses" away.

The two scientists spearheading the research, Dr. Paula Tallal of Rutgers University in Newark and Dr. Michael Merzenich of the University of California School of Medicine in San Francisco, said in interviews that they believed that the treatment would help many children and adults with milder forms of language and reading disability—the condition widely known as dyslexia. But Dr. Tallal, who is director of the Center for Molecular and Behavioral Neuroscience at the Newark campus, and Dr. Merzenich, who is a professor of otolaryngology and physiology, cautioned that dyslexia had numerous causes and that not everyone with reading problems would respond to the treatment.

Their new findings, along with the first detailed description of the treatment, have just been submitted for publication in a leading scientific journal. The researchers declined to disclose the precise contents of the journal article. But they have talked about their work at several scientific meetings this year and presented results in San Diego at the annual meeting of the Society for Neuroscience.

Dr. Sally Shaywitz, a leading expert on dyslexia at Yale University, heard an oral presentation of the research two weeks ago and said the findings were "tremendously exciting." But, she added, "I am not convinced" that the majority of dyslexic children will be helped by these methods. "I've seen many things with promise over the years fail to deliver," Dr. Shaywitz said, adding that "while this work sounds credible, we need more experiments" before parents and teachers get their hopes aroused.

But other experts are more optimistic. "I think Paula Tallal and Mike Merzenich's work is just superb," said Dr. Ursula Bellugi, director of the Laboratory for Cognitive Neuroscience at the Salk Institute in San Diego. She said it was possibly the first therapy for dyslexia based on a deeper understanding of the way the brain is organized.

Ten million American children suffer from dyslexia, defined as having great difficulty in reading single words despite normal intelligence and motivation. The social costs are tremendous. A high percentage of dyslexic children drop out of school and have substance abuse problems. They also tend to have trouble finding jobs.

Dr. Tallal is an expert on language impairments in children. Dr. Merzenich is an authority on brain plasticity—how brain cells and cortical maps change in response to experience. The two began collaborating a couple of years ago under a grant from the Charles A. Dana Foundation in New York.

Dr. Tallal's research focuses on children with a condition called specific language impairment. These children do not talk normally by the age of 3 or 4, Dr. Tallal said. They have trouble hearing and generating speech—they might say titty tat for kitty cat—and often cannot follow directions from parents and teachers.

Four in five of these children go on to become severely dyslexic in grade school, Dr. Tallal said. But they are a small subset of the larger population of children with reading problems.

Nevertheless, Dr. Tallal said she believed that these children provided insights into the wider problem. And she thinks she knows what it is.

Research shows that tiny infants can discriminate the sounds in all spoken languages, Dr. Tallal said. But by 6 months of age, if their hearing is normal, they begin to extract sounds that are salient in their native tongue. Specifically, they focus on phonemes—the basic sound units of a language—and begin to practice them by babbling. English has 44 phonemes—sounds like bah, dee and moo—that can be combined to make up the hundreds of thousands of words in the English language.

Given this normal process, Dr. Tallal decided to take a closer look at phonemes and the way children learn language. Some sounds, such as pure vowels like aaaahhh, occur in a steady flow that continues for more than 100 milliseconds (a tenth of a second), she said. But other sounds, such as ba, da, ga, pa, ta and ka, have a different spectral shape. For example, in the phoneme "ba" the initial "b" is formed by pressing the lips together silently. But then there is a rapid transition, lasting 40 milliseconds, to the "aaah" sound, she said.

The brain has to distinguish these fast transitions to discriminate "ba" from "da" or "ta" from "ka," Dr. Tallal said. It must be able to detect frequency changes in four tenths of a second.

Other phonemes do not require a fast transition, she said. The initial "m" in the phoneme "ma" is formed by pressing the lips together and voicing the sound. There is a slow transition lasting perhaps 300 milliseconds before the "aaah" is voiced.

These slow and rapid sound transitions carry meaning, Dr. Tallal said. It is the glue that holds speech together.

Dr. Tallal suspects that this is where the problem lies in language-impaired children. For reasons that are not yet well understood, cells in their primary auditory cortex cannot detect rapid transition phonemes, she said. In some cases, the problem could be genetic. In others, chronic ear infections in infancy might cause the auditory cortex to miss important cues during a critical period of development; cells could become tuned to slower but not faster frequencies.

A child with this problem cannot hear the difference between "ba" and "da," Dr. Tallal said. Certain phonemes are eternally garbled, while others make sense.

Most children learn to compensate, Dr. Tallal said. There is a lot of redundancy in spoken language, so they can listen for words they do understand. They pay attention to body language and facial expressions. And they pull meaning from context. For example, "the man chased the dog" and "the man chased the bog" would sound similar but the child knows, from common sense, that the first sentence carries the intended meaning, Dr. Tallal said.

Some language-impaired children appear to speak normally by the age of 5 or 6 but they are faking it, Dr. Tallal said. They get the gist of many sentences but not all the meaning.

Others live in a language fog, she said. "It's like you having a minimal ability in a foreign language and visiting that country," she said. "All day long you struggle." Boys tend to act out and become aggressive or hyperactive. "If someone spoke mumbo jumbo to you all day long, would you be able to sit still?" she asked.

Then the children hit first grade, Dr. Tallal said. Teachers are under the impression that speech has been learned automatically, before children come to school. And so they teach reading by phonics, featuring the very phonemes that language-impaired children cannot fathom.

When Dr. Merzenich, an expert on the plasticity of brain organization, met Dr. Tallal, she told him about the problem some children had in hearing fast transition sounds. The two thought that perhaps they could force changes in the human auditory cortex. Perhaps they could "wake up" sluggish cells to help children hear fast transition phonemes.

"We decided to give kids information in a way their brains can process it," Dr. Merzenich said. The key ingredient is "processed speech" generated by a computer. Fast transition phonemes are stretched out artificially. Instead of the natural 40 milliseconds between "b" and "ah," the computer generates "ba" with 300, 400 or even 500 milliseconds between "b" and "ah." The computer also emphasizes difficult-to-hear phonemes, making them louder, longer and more salient to the child's brain.

To a normal adult, the processed speech sounds like someone shouting underwater. To language-impaired children, it is miraculous, the researchers said. The children can hear many words clearly for the first time in their lives.

The new treatment exploits what most parents already know—namely, that children will sit for hours in front of a video game. Dr. William Jenkins at the San Francisco campus developed four colorful computer games with processed speech. The games drill children in hearing pairs of tones and phonemes at faster and faster rates of speed.

The training is done with flair and fanfare. Cows on rocket ships float across the screen. Clowns appear and make wild noises. Bells, flashing lights and other surprises appear when children make high scores.

An important feature is that the games are driven by the child's own performance, Dr. Merzenich said. For example, when children show they can reliably hear the difference between "ba" and "pa"—processed with say 300 milliseconds of transition time—the transition is shortened, to say 280 milliseconds. If the child makes errors, the time is lengthened back to 300 milliseconds and drilling continues. Whenever a level is mastered, the time is shortened, thus driving the child's auditory cortex to handle faster and faster sounds.

Children practice thousands of sounds each session, always returning to the level they achieved on the previous day, Dr. Merzenich said. Processed speech is the grammar they can use to win points in the computer games, he said, and they play to win.

Processed speech, played on a tape recorder, is also used in a variety of one-on-one exercises with the children. They learn to listen to speech, attend to grammar and follow directions.

Finally, the children take home books on tape, like *Cat in the Hat,* recorded in processed speech.

Two summers ago, Dr. Tallal and Dr. Merzenich gave their experimental treatment to seven language-impaired children ranging in age from 5 to 9. The children were at least two years behind in language skills, Dr. Tallal said. Their speech was often garbled.

Children visited a laboratory at Rutgers for three and a half hours a day, five days a week, for six weeks. Their language and reading comprehension were tested the first week. The therapy was given for four weeks. Then the children were re-tested on the sixth week.

These children made two years of progress in just one month, Dr. Tallal said. After the therapy, they were performing at or a little above their age level in receptive grammar—the ability to comprehend spoken words.

Three months later the children were tested again. They had not slid backward. Evidently, natural speech served to reinforce the gains they had made in the laboratory, Dr. Merzenich said.

Amazed by these results, Dr. Tallal and Dr. Merzenich held off from submitting a scientific article. Valid questions came up, Dr. Tallal said. "Maybe just the intense intervention, that warm and cuddling feeling, made the difference," she said. "When someone is constantly telling you you're good, doing a good job, you may do better. Maybe we just improved their memories."

To control those variables, the researchers repeated their experiment this summer with 22 similar children. Half got all the computer games and one-on-one exercises with processed speech. The other half got the same treatment, including computer games, but without processed speech.

Results were again striking. While all the children improved, those exposed to the processed speech outperformed the others—achieving two years' improvement after one month of therapy.

These gains should help the children with their reading, Dr. Tallal said. But long-term follow-up studies have yet to be done.

"We really don't know how far we can drive these kids to improve," Dr. Merzenich said. One month's therapy is probably not enough, he said. "They've had a lifetime of practicing the wrong speech sounds, but there's nothing to indicate that we can't change that."

A major question remains. Do these findings on severely language-impaired children apply to the hundreds of thousands of people diagnosed with dyslexia?

Dr. Shaywitz, who is carrying out an epidemiological study on dyslexia, said that the condition was like hypertension or obesity. It seems to fall on a continuum from mild to severe. While some children cope, the deficit does not go away. It is not something they simply outgrow, she said.

But it is not yet known if language-impaired children are on a continuum with dyslexia, Dr. Shaywitz said.

Dr. Tallal agreed and said that more studies were needed. There are certainly children who have speech problems that are not related to temporal coding, she said. "But my hunch is that the vast majority of kids with reading problems have phonological access problems," she said.

—SANDRA BLAKESLEE, November 1995

Linguists Debate Study Classifying Language as Innate Human Skill

A STUDY of a deaf child's linguistic abilities is stirring up an ancient debate over the nature of language. Is the human brain uniquely programmed to make and learn languages or does it simply pick up on ordered structures perceived when a child is first exposed to speech?

The subject, a 9-year-old boy named Simon, is uniquely appropriate for the experiment of asking whether language is learned or innate because he learned an error-riddled form of American Sign Language from his parents, who are also deaf, and a quite different sign language, with different grammatical rules, at his school. Despite the faulty teaching of American Sign Language, Simon signed the language with correct grammar, which the researchers see as evidence that he was drawing upon innate language ability.

The researchers studied Simon from the time he was 2½ years old to the time he was 9. They reported that he had signed in American Sign Language with the correct grammar, even though he had learned incorrect grammar from his parents.

To the obvious objection that Simon may have seen other people signing correctly in American Sign Language, the researchers reply that his parents were the only people whom he had seen signing in American Sign Language, apart from his parents' friends, who also signed incorrectly. His parents and their friends learned American Sign Language only as teenagers, an age at which languages are often learned inaccurately.

The investigators, Dr. Elissa L. Newport of the University of Rochester and Dr. Jenny L. Singleton of the University of Illinois, believe that Simon recognized complex patterns in the language on the basis of his parents' inconsistent use of the patterns. And, they say, Simon learned to use some complicated rules in ways that had eluded his parents. The research was presented at a meeting of the American Psychological Society in San Diego.

Dr. Newport said that the way Simon had deduced grammatical rules showed "exactly the kinds of things you would predict" from theories of how children develop language. But it has been very difficult to find evidence that these theories are correct.

Other investigators said they were intrigued by this case. Dr. Ursula Bellugi, a neuroscientist at the Salk Institute in San Diego, California, said, "It has been hard to get really solid evidence of whether the brain is disposed in particular ways for learning languages." The story of Simon, she said, "is really exciting" because it is so scientifically clean. "I think it's very convincing," she added.

And Dr. Susan Goldin-Meadow of the University of Chicago said, "I think their data are incontrovertible."

But Dr. Jean Berko Gleason of Boston University, who is the editor of the standard linguistics textbook *The Development of Language,* said that she would not read so much into the case history. Simon, she said, "seemed to pick up on the regularities of the language," but he did not invent language structure out of whole cloth. And the study is based on just a single child, she added. "It's always interesting even if one child does something, but you never know if he's showing universal tendencies," Dr. Berko Gleason said.

Simon's story is part of a centuries-long tradition of case studies of children who scientists hoped could help shed light on the question of whether language is innate and whether there is only a window of time, when children are maturing, in which it can be learned. Researchers have studied feral children, who are called that because they grew up with only animals for company. They have studied abused children who had been kept isolated and deprived of human talk and companionship. They have studied deaf children who had not been taught to sign.

But these studies were not scientifically pure, researchers said. The feral children and abused children had so many other emotional and physical problems that it was impossible to say what was cause and what was effect. The deaf children developed a language so simple that some question whether it counts.

Simon, on the other hand, was loved and cherished and was taught a language by his parents. The only thing missing was consistently correct complex grammar and sentence structure.

Although Simon's parents were each born deaf to hearing parents, they did not learn to sign as children. Instead, they were sent to schools that tried to teach them to read lips. Like most deaf people, they never succeeded in this endeavor. Only as teenagers did they learn American Sign Language, and they learned it imperfectly.

But, the researchers said, Simon divined grammatical rules that his parents could not grasp. The parents, for example, had trouble with verbs of motion, which they used correctly just 65 percent of the time; Simon signed the verbs correctly 90 percent of the time.

Simon's parents had no grasp of some grammatical rules of American Sign Language, like a rule called topicalization, which allows the signer to make a word the topic of a sentence even though it is not the subject. With topicalization, Dr. Newport said, "You can say, 'John hit Mary,' but if you want to talk about Mary, you would say, 'Mary, John hit.'" While moving the word "Mary" to the front of the phrase, the signer makes a special facial expression, with lifted eyebrows and chin and drawn-up muscles under the nose.

Although Simon's parents never moved words in sentences, they would emphasize the first word in a sentence by making the special facial expression. "Simon's parents don't seem to know the movement rule," Dr. Newport said. But Simon, on the other hand, "does it perfectly correctly." Since Simon never saw the rule used correctly, "I don't think he's extracting it from the pattern of input. I think it's something he's born with," she added.

"There is a very rich argument that kids must somehow be equipped with a lot of biases that make them organize languages in particular ways," Dr. Newport said.

For example, Dr. Noam Chomsky of the Massachusetts Institute of Technology, who initiated the modern era in linguistics in the late 1950s with his studies of language structure, has argued that all the world's languages share common features that reflect a biological determinism. He believes that all children are surrounded by errors and incompleteness when learning language but that they pick out the rich grammatical structures, developing a grasp of language that goes beyond their exposure. "What they know is so far beyond what they've heard that they obviously created it themselves," he said.

In other studies, Dr. Goldin-Meadow and her colleagues examined the deaf children of hearing parents who had not been taught any sign language and asked whether they had made up a language for themselves. About 20 years ago, it was common for deaf children to be discouraged from learning to sign, Dr. Goldin-Meadow said. "The parents don't know sign language, and they want their child to be part of the hearing world. If the child learns sign language, they will eventually leave that world," she said. So the parents and schools tried to teach the children to read lips and to speak. "Very, very, very few children ever succeeded, but there was always that hope," Dr. Goldin-Meadow said.

But, in the meantime, the children would make up their own signs. The language was simple, involving pointing and gesturing. But, Dr. Goldin-Meadow found, the children would start to string gestures together, while the parents would only rarely elaborate on the signs. A child, for example, "would point at a cup and then make gestures for drinking. But a parent would only rarely combine gestures, and even if they did combine them they would not combine them with such order," Dr. Goldin-Meadow said.

Still other studies, by Dr. Bellugi, involve deaf people whose brains had been injured by strokes. She discovered that the same area of the left brain is used for all language, whether spoken or signed, even though sign language is such a spatial language. One woman studied by Dr. Bellugi had suffered a stroke that had injured the right hemisphere of her brain.

"She was an artist and she was unable to draw; she couldn't do perspective," Dr. Bellugi said. "But her signing was perfect." Dr. Bellugi said she was convinced that "language acquisition is very much biologically determined."

But Dr. Berko Gleason said that these studies, while intriguing, still left open the question of whether the capacity to learn language is innate. Although the deaf children invent a sign language when they are not taught to sign, she said, "whether that is language as we know it is open to question." And the brain-injury studies do not prove that the language area in the brain was there before people learned language. With virtually all the studies, Dr. Berko Gleason said, "it is always the same problem: there are very small samples, and things are very much in the eye of the beholder."

But Dr. Chomsky disagrees. He said the idea that the human brain is organized to make the learning of language innate "is strongly established." Although he did not need Simon to convince him, he said, Simon's case shows that children extract more from language than they are ever explicitly shown.

—GINA KOLATA, September 1992

Chimp Talk Debate: Is It Really Language?

PANBANISHA, a Bonobo chimpanzee who has become something of a star among animal language researchers, was strolling through the Georgia woods with a group of her fellow primates—scientists at the Language Research Center at Georgia State University in Atlanta. Suddenly, the chimp pulled one of them aside. Grabbing a special keyboard of the kind used to teach severely retarded children to communicate, she repeatedly pressed three symbols—"Fight," "Mad," "Austin"—in various combinations.

Austin is the name of another chimpanzee at the center. Dr. Sue Savage-Rumbaugh, one of Panbanisha's trainers, asked, "Was there a fight at Austin's house?"

"Waa, waa, waa," said the chimpanzee, in what Dr. Savage-Rumbaugh took as a sign of affirmation. She rushed to the building where Austin lives and learned that earlier in the day two of the chimps there, a mother and her son, had fought over which got to play with a computer and joystick used as part of the training program. The son had bitten his mother, causing a ruckus that, Dr. Savage-Rumbaugh surmised, had been overheard by Panbanisha, who lived in another building about 200 feet away. As Dr. Savage-Rumbaugh saw it, Panbanisha had a secret she urgently wanted to tell.

A decade and a half after the claims of animal language researchers were discredited as exaggerated self-delusions, Dr. Savage-Rumbaugh is reporting that her chimpanzees can demonstrate the rudimentary comprehension skills of 2½-year-old children. According to a series of recent papers, the Bonobo, or pygmy, chimps, which some scientists believe are more human-like and intelligent than the common chimpanzees studied in the earlier, flawed experiments, have learned to understand complex sentences and use symbolic language to communicate spontaneously with the outside world.

"She had never put those three lexigrams together," Dr. Savage-Rumbaugh said, referring to the keyboard symbols with which the animals are trained. She found the incident, which occurred last month, particularly gratifying because the chimp seemed to be using the symbols not to demand food, which is usually the case in these experiments, but to gossip.

In their book *Apes, Language and the Human Mind: Philosophical Primatology* (Routledge), Dr. Savage-Rumbaugh and her co-authors, Dr. Stuart Shanker, a philosopher at York University in Toronto, and Dr. Talbot Taylor, a linguist at the College of William and Mary in Virginia, argue that the feats of the chimps at the Language Research Center are so impressive that scientists must now re-evaluate some of their most basic ideas about the nature of language.

Most language experts dismiss experiments like the ones with Panbanisha as exercises in wishful thinking. "In my mind this kind of research is more analogous to the bears in the Moscow circus who are trained to ride unicycles," said Dr. Steven Pinker, a cognitive scientist at the Massachusetts Institute of Technology who studies language acquisition in children. "You can train animals to do all kinds of amazing things." He is not convinced that the chimps have learned anything more sophisticated than how to press the right buttons in order to get the hairless apes on the other side of the console to cough up M & M's, bananas and other tidbits of food.

Dr. Noam Chomsky, the MIT linguist whose theory that language is innate and unique to people forms the infrastructure of the field, says that attempting to teach linguistic skills to animals is irrational—like trying to teach people to flap their arms and fly.

"Humans can fly about 30 feet—that's what they do in the Olympics," he said in an interview. "Is that flying? The question is totally meaningless. In fact the analogy to flying is misleading because when humans fly 30 feet, the organs they're using are kind of homologous to the ones that chickens and eagles use." Arms and wings, in other words, arise from the same branch of the evolutionary tree. "Whatever the chimps are doing is not even homologous as far as we know," he said. There is no evidence that the chimpanzee utterances emerge from anything like the "language organ" Dr. Chomsky believes resides only in human brains. This neural wiring is said to be the source of the universal grammar that unites all languages.

But some philosophers, like Dr. Shanker, complain that the linguists are applying a double standard: They dismiss skills—like putting together a noun and a verb to form a two-word sentence—that they consider nascent linguistic abilities in a very young child.

"The linguists kept upping their demands and Sue kept meeting the demands," said Dr. Shanker. "But the linguists keep moving the goal post."

Following Dr. Chomsky, most linguists argue that special neural circuitry needed for language evolved after man's ancestors split from those of the chimps millions of years ago. As evidence they note how quickly children, unlike chimpanzees, go from cobbling together two-word utterances to effortlessly spinning out complex sentences with phrases embedded within phrases like Russian dolls. But Dr. Shanker and his colleagues insist that Dr. Savage-Rumbaugh's experiments suggest that there is not an unbridgeable divide between humans and the rest of the animal kingdom, as orthodox linguists believe, but rather a gradation of linguistic skills.

In a book, *The Engine of Reason, the Seat of the Soul: A Philosophical Journey Into the Brain* (MIT Press), Dr. Paul Churchland, a philosopher and cognitive scientist at the University of California at San Diego, says linguists should take Dr. Savage-Rumbaugh's experiments as a challenge. He argues that the jury is still out: The rules for constructing sentences might turn out to be not so much hard-wired as a result of learning—by people and potentially by their chimpanzee relatives.

Animal language research fell into disrepute in the late 1970s when "talking" chimps like Washoe and the provocatively named Nim Chimpsky were exposed as unintentional frauds. Because chimpanzees lack the vocal apparatus to make a variety of modulated sounds, the animals were taught a vocabulary of hand signs—an approach first suggested in the 18th century by the French physician Julien Offray de La Mettrie. In appearances on television talk shows, trainers claimed the chimps could construct sentences of several words. But upon closer examination, scientists found strong evidence that the chimps had simply learned to please their teachers by contorting their hands into all kinds of configurations. And the trainers, straining to find examples of linguistic communication, thought they saw words among the wiggling, like children seeing pictures in the clouds.

In a widely quoted paper in the journal *Science*, "Can an Ape Create a

Sentence?" Nim Chimpsky's trainer, Dr. Herbert Terrace, a Columbia University psychologist, reluctantly concluded that the answer was no.

A chimp might learn to connect a hand sign with an item of food, skeptics like Dr. Terrace argued, but this could be a matter of simple conditioning, like Pavlov's dogs learning to salivate at the sound of a bell. Most importantly, there was no evidence that the chimps had acquired a generative grammar—the ability to string words together into sentences of arbitrary length and complexity.

As a young veteran of the original animal language experiments, Dr. Savage-Rumbaugh decided to try a different approach. To eliminate the ambiguity of hand signs, she used a keyboard with dozens of buttons marked with geometric symbols.

In elaborate exercises beginning in the mid-1970s, she and her colleagues taught common chimpanzees and Bonobos to associate symbols with a variety of things, people and places in and around the laboratory. The smartest chimps even seemed to learn abstract categories, identifying pictures of objects as either tools or food. Dr. Savage-Rumbaugh reported that two of the chimps learned to use symbols to communicate with each other. Pecking away at the keyboard, one would tell a companion where to find a key that would liberate a banana for them both to share.

Most impressive of all was a Bonobo named Kanzi. After futilely trying to train Kanzi's adopted mother to use the keyboard, the researchers found that the 2½-year-old chimp, who apparently had been eavesdropping all along, had picked up an impressive vocabulary on his own. Kanzi was taught not in laboriously structured training sessions but on walks through the 50 acres of forest surrounding the language center. By the time he was 6 years old, Kanzi had acquired a vocabulary of 200 symbols and was constructing what might be taken as rudimentary sentences consisting of a word combined with a gesture or occasionally of two or three words. Dr. Savage-Rumbaugh became convinced that exposure to language must start early and that the lessons should be driven by the animal's curiosity.

Compared with other chimps, Kanzi's utterances are striking, but they are still far from human abilities. Kanzi is much better at responding to vocal commands like "Take off Sue's shoe." In one particularly arresting feat, recorded on videotape, Kanzi was told,"Give the dog a shot." The chimpanzee picked up a hypodermic syringe lying on the ground in front of him, pulled off the cap and injected a toy stuffed dog.

Dr. Savage-Rumbaugh's critics say there is nothing surprising about chimpanzees or even dogs and parrots associating vocal sounds with objects. Kanzi has been trained to associate the sound "dog" with the furry thing in front of him and has been programmed to carry out a stylized routine when he hears "shot." But does the chimp really understand what he is doing?

Dr. Savage-Rumbaugh insists that experiments using words in novel contexts show that the chimps are not just responding to sounds in a knee-jerk manner. It is true, she says, that Kanzi was initially aided by vocal inflections, hand gestures, facial expressions and other contextual clues. But once it had mastered a vocabulary, the Bonobo could properly respond to 70 percent of unfamiliar sentences spoken by a trainer whose face was concealed.

None of this is very persuasive to linguists for whom the acid test of language is not comprehension but performance, the ability to use grammar to generate ever more complex sentences.

Dr. Terrace says Kanzi, like the disappointing Nim Chimpsky, is simply "going through a bag of tricks in order to get things." He is not impressed by comparisons to human children. "If a child did exactly what the best chimpanzee did, the child would be thought of as disturbed," Dr. Terrace said.

The scientists at the Language Research Center are "studying some very complicated cognitive processes in chimpanzees," Dr. Terrace said. "That says an awful lot about the evolution of intelligence. How do chimpanzees think without language, how do they remember without language? Those are much more important questions than trying to reproduce a few tidbits of language from a chimpanzee trying to get rewards."

Attempting to shift the fulcrum of the debate over performance versus comprehension, Dr. Savage-Rumbaugh argues that the linguists have things backward: "Comprehension is the route into language," she says. In her view it is easier to take an idea already in one's mind and translate it into a grammatical string of words than to decipher a sentence spoken by another whose intentions are unknown.

Dr. Shanker, the York University philosopher, believes that the linguists' objections reveal a naive view of how language works. When Kanzi gives the dog a shot, he might well be relying on all kinds of contextual clues and subtle gestures from the speaker, but that, Dr. Shanker argues, is what people do all the time.

Following the ideas of the philosopher Ludwig Wittgenstein, he argues that language is not just a matter of encoding and decoding strings of arbitrary symbols. It is a social act that is always embedded in a situation.

But trotting out Wittgenstein and his often obscure philosophy is a way of sending many linguists bolting for the exits. "If higher apes were incapable of anything beyond the trivialities that have been shown in these experiments, they would have been extinct millions of years ago," Dr. Chomsky said. "If you want to find out about an organism you study what it's good at. If you want to study humans you study language. If you want to study pigeons you study their homing instinct. Every biologist knows this. This research is just some kind of fanaticism."

There is a suspicion among some linguists and cognitive scientists that animal language experiments are motivated as much by ideological as scientific concerns—by the conviction that intelligent behavior is not hard-wired but learnable, by the desire to knock people off their self-appointed thrones and champion the rights of downtrodden animals.

"I know what it's like," Dr. Terrace said. "I was once stung by the same bug. I really wanted to communicate with a chimpanzee and find out what the world looks like from a chimpanzee's point of view."

—GEORGE JOHNSON, June 1995

5

GROWING A MIND

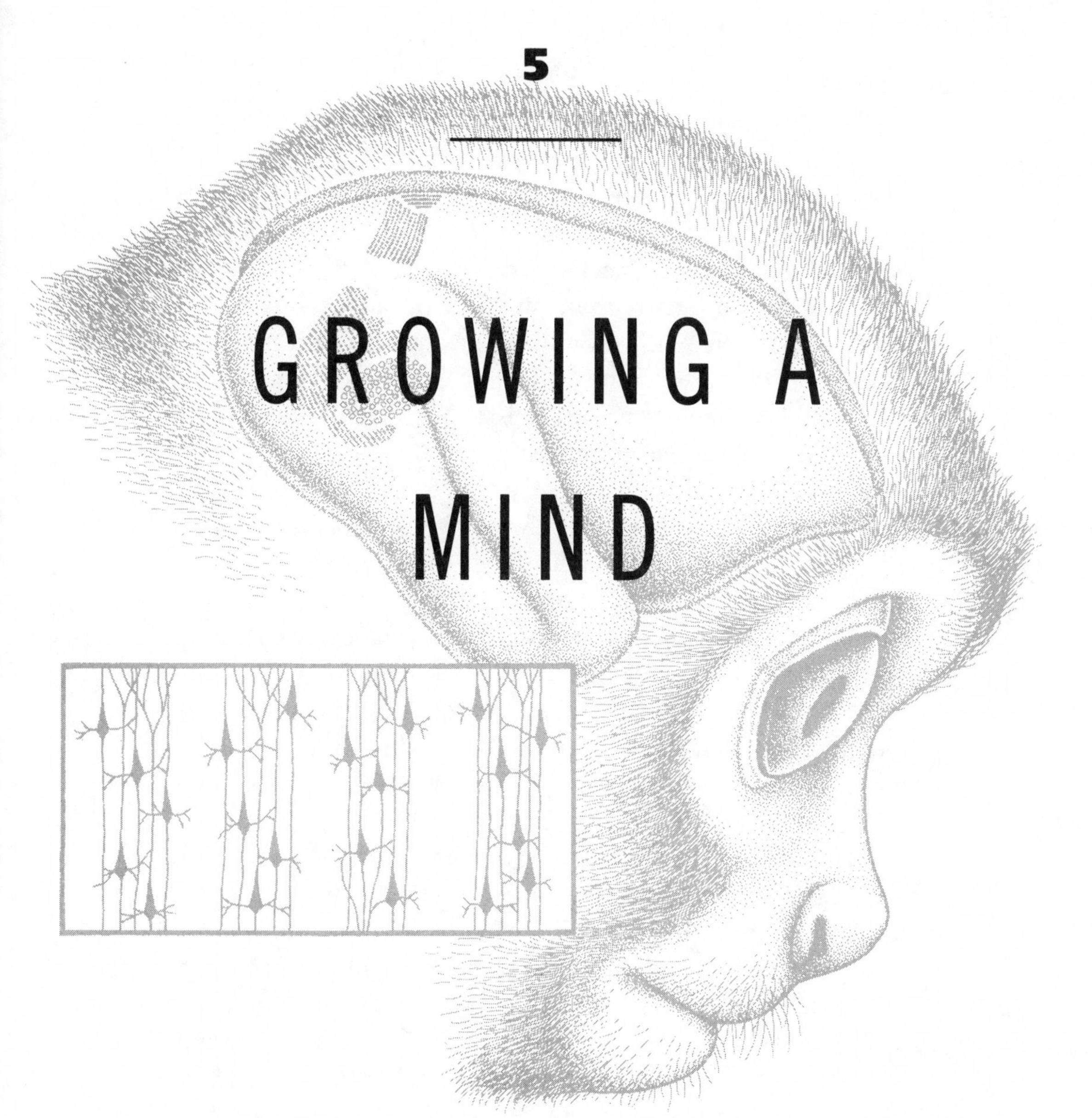

The human brain holds about 100 billion nerve cells, or neurons. Each neuron forms contacts with a thousand others, on average, making for a system with at least 100 trillion interconnections. Moreover the system appears to be wired up with almost perfect accuracy.

How is this rather impressive feat of engineering accomplished? The basic wiring plan must be encoded in the genes. It is evident from studying the developing embryos of animals that nerve cells are created in or travel to designated regions of the brain, and once in place send out axons along pre-programmed paths to make contact with specific targets.

Neurobiologists have recently discovered another basic principle, one that perhaps explains the extraordinary fidelity of the brain's interconnections. The basic wiring plan put in place by genetic instructions becomes modifiable by experience during a critical interval in the infant's life. Axons from the retina of the eye grow into the developing brain and make a multitude of connections in the optic cortex. There is then a kind of testing period during which the connections that are less used get pruned away in favor of those that make the system work right.

This refinement of the wiring system may be a general principle of the developing brain. It may also be the mechanism that underlies the existence of critical windows of time during which an infant must learn certain skills, such as the period in about the fourth month of life when the infant's brain develops binocular vision. There seem to be other critical periods, less precisely defined, for the development of language and sociability. Children who for some reason fail to acquire these skills during the critical period may never do so.

As the following articles show, the development of the brain is another topic in which some of the general principles are emerging but the details are far from being fully understood.

Studies Show Talking with Infants Shapes Basis of Ability to Think

THE NEUROLOGICAL FOUNDATIONS for rational thinking, problem solving and general reasoning appear to be largely established by age 1—long before babies show any signs of knowing an abstraction from a pacifier.

Furthermore, new studies are showing that spoken language has an astonishing impact on an infant's brain development. In fact, some researchers say the number of words an infant hears each day is the single most important predictor of later intelligence, school success and social competence. There is one catch—the words have to come from an attentive, engaged human being. As far as anyone has been able to determine, radio and television do not work.

"We now know that neural connections are formed very early in life and that the infant's brain is literally waiting for experiences to determine how connections are made," said Dr. Patricia Kuhl, a neuroscientist at the University of Washington in Seattle and a key speaker at today's conference. "We didn't realize until very recently how early this process begins," she said in a telephone interview. "For example, infants have learned the sounds of their native language by the age of six months."

This relatively new view of infant brain development, supported by many scientists, has obvious political and social implications. It suggests that infants and babies develop most rapidly with caretakers who are not only loving, but also talkative and articulate, and that a more verbal family will increase an infant's chances for success. It challenges some deeply held beliefs—that infants will thrive intellectually if they are simply given lots of love and that purposeful efforts to influence babies' cognitive development are harmful.

If the period from birth to 3 is crucial, parents may assume a more crucial role in a child's intellectual development than teachers, an idea sure to provoke new debates about parental responsibility, said Dr. Irving Lazar, a professor of special education and resident scholar at the Center for Research in Human Development at Vanderbilt University in Nashville. And it offers yet another reason to provide stimulating, high-quality day care for infants whose primary caretakers work, which is unavoidably expensive.

The idea that early experience shapes human potential is not new, said Dr. Harry Chugani, a pediatric neurologist now at Wayne State University in Detroit and one of the scientists whose research has shed light on critical periods in child brain development. What is new is the extent of the research in the field known as cognitive neuroscience and the resulting synthesis of findings on the influence of both nature and nurture. Before birth, it appears that genes predominantly direct how the brain establishes basic wiring patterns. Neurons grow and travel into distinct neighborhoods, awaiting further instructions.

After birth, it seems that environmental factors predominate. A recent study found that mice exposed to an enriched environment have more brain cells than mice raised in less intellectually stimulating conditions. In humans, the inflowing stream of sights, sounds, noises, smells, touches—and most importantly, language and eye contact—literally makes the brain take shape. It is a radical and shocking concept.

Experience in the first year of life lays the basis for networks of neurons that enable us to be smart, creative and adaptable in all the years that follow, said Dr. Esther Thelen, a neurobiologist at Indiana University in Bloomington.

The brain is a self-organizing system, Dr. Thelen said, with many parts that co-operate to produce coherent behavior. There is no master program pulling it together. "What we know about these systems is that they are very sensitive to initial conditions," Dr. Thelen said. "Where you are now depends on where you've been."

The implication for infant development is clear. Given the explosive growth and self-organizing capacity of the brain in the first year of life, the experiences an infant has during this period are the conditions that set the stage for everything that follows.

In later life, what make us smart and creative and adaptable are networks of neurons which support our ability to use abstractions from one memory to help form new ideas and solve problems, said Dr. Charles Stevens, a neurobiologist at the Salk Institute in San Diego. Smarter people may have a greater number of neural networks that are more intricately woven together, a process that starts in the first year.

The complexity of the synaptic web laid down early may very well be the physical basis of what we call general intelligence, said Dr. Lazar at Vanderbilt. The more complex that set of interconnections, the brighter the child is likely to be since there are more ways to sort, file and access experiences.

Of course, brain development "happens" in stimulating and dull environments. Virtually all babies learn to sit up, crawl, walk, talk, eat independently and make transactions with others, said Dr. Steve Petersen, a neurologist at Washington University School of Medicine in St. Louis. Such skills are not at risk except in rare circumstances of sensory and social deprivation, like being locked in a closet for the first few years of life. Subject to tremendous variability within the normal range of environments are the abilities to perceive, conceptualize, understand, reason, associate and judge. The ability to function in a technologically complex society like ours does not simply "happen."

One implication of the new knowledge about infant brain development is that intervention programs like Head Start may be too little, too late, Dr. Lazar said. If educators hope to make a big difference, he said, they will need to develop programs for children from birth to 3.

Dr. Betty Caldwell, a professor of pediatrics and an expert in child development at the University of Arkansas in Little Rock, who supports the importance of early stimulation, said that in early childhood education there is a strong bias against planned intellectual stimulation. Teachers of very young children are taught to follow "developmentally appropriate practices," she said, which means that the child chooses what he or she wants to do. The teacher is a responder and not a stimulator.

Asked about the bias Dr. Caldwell described, Matthew Melmed, executive director of Zero to Three, a research and training organization for early childhood development in Washington, D.C., said that knowing how much stimulation is too much or too little, especially for infants, is "a really tricky question. It's a dilemma parents and educators face every day," he said.

In a poll released today, Zero to Three found that 87 percent of parents think that the more stimulation a baby receives the better off the baby is,

Mr. Melmed said. "Many parents have the concept that a baby is something you fill up with information and that's not good," he said.

"We are concerned that many parents are going to take this new information about brain research and rush to do more things with their babies, more activities, forgetting that it's not the activities that are important. The most important thing is connecting with the baby and creating an emotional bond," Mr. Melmed said.

There is some danger of overstimulating an infant, said Dr. William Staso, a school psychologist from Orcutt, California, who has written a book called *What Stimulation Your Baby Needs to Become Smart.* Some people think that any interaction with very young children that involves their intelligence must also involve pushing them to excel, he said. But the "curriculum" that most benefits young babies is simply common sense, Dr. Staso said. It does not involve teaching several languages or numerical concepts but rather carrying out an ongoing dialogue with adult speech. Vocabulary words are a magnet for a child's thinking and reasoning skills.

This constant patter may be the single most important factor in early brain development, said Dr. Betty Hart, a professor emeritus of human development at the University of Kansas in Lawrence. With her colleague, Dr. Todd Ridley of the University of Alaska, Dr. Hart recently co-authored a book—*Meaningful Differences in the Everyday Experience of Young American Children.*

The researchers studied 42 children born to professional, working-class or welfare parents. During the first two and half years of the children's lives, the scientists spent an hour a month recording every spoken word and every parent-child interaction in every home. For all the families, the data include 1,300 hours of everyday interactions, Dr. Hart said, involving millions of ordinary utterances.

At age 3, the children were given standard tests. The children of professional parents scored highest. Spoken language was the key variable, Dr. Hart said.

A child with professional parents heard, on average, 2,100 words an hour. Children of working-class parents heard 1,200 words and those with parents on welfare heard only 600 words an hour. Professional parents talked three times as much to their infants, Dr. Hart said. Moreover, children with professional parents got positive feedback 30 times an hour—twice as often as working-class parents and five times as often as welfare parents.

The tone of voice made a difference, Dr. Hart said. Affirmative feedback is very important. A child who hears, "What did we do yesterday? What did we see?" will listen more to a parent than will a child who always hears "Stop that," or "Come here!"

By age 2, all parents started talking more to their children, Dr. Hart said. But by age 2, the differences among children were so great that those left behind could never catch up. The differences in academic achievement remained in each group through primary school.

Every child learned to use language and could say complex sentences but the deprived children did not deal with words in a conceptual manner, she said.

A recent study of day care found the same thing. Children who were talked to at very young ages were better at problem solving later on.

For an infant, Dr. Hart said, all words are novel and worth learning. The key to brain development seems to be the rate of early learning—not so much what is wired but how much of the brain gets interconnected in those first months and years.

TIMETABLE: The Growing Brain: What Might Help Your Infant

Dr. William Staso, an expert in neurological development, suggests that different kinds of stimulation should be emphasized at different ages. At all stages, parental interaction and a conversational dialogue with the child are important. Here are some examples:

FIRST MONTH—A low level of stimulation reduces stress and increases the infant's wakefulness and alertness. The brain essentially shuts down the system when there is overstimulation from competing sources. When talking to an infant, for example, filter out distracting noises, like a radio.

MONTHS 1 TO 3—Light/dark contours, like high-contrast pictures or objects, foster development in neural networks that encode vision. The brain also starts to discriminate among acoustic patterns of language, like intonation, lilt and pitch. Speaking to the infant, especially in an animated voice, aids this process.

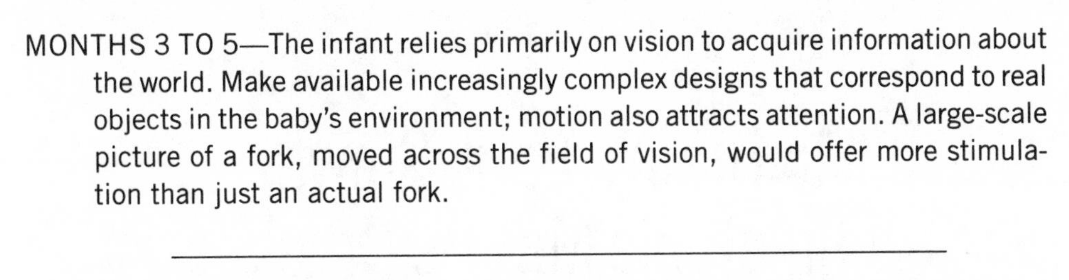

MONTHS 3 TO 5—The infant relies primarily on vision to acquire information about the world. Make available increasingly complex designs that correspond to real objects in the baby's environment; motion also attracts attention. A large-scale picture of a fork, moved across the field of vision, would offer more stimulation than just an actual fork.

MONTHS 6 TO 7—The infant becomes alert to relationships like cause and effect, the location of objects and the functions of objects. Demonstrate and talk about situations like how the turning of a doorknob leads to the opening of a door.

MONTHS 7 TO 8—The brain is oriented to make associations between sounds and some meaningful activity or object. For example, parents can deliberately emphasize in conversation that the sound of water running in the bathroom signals an impending bath, or that a doorbell means a visitor.

MONTHS 9 TO 12—Learning adds up to a new level of awareness of the environment and increased interest in exploration; sensory and motor skills coordinate in a more mature fashion. This is the time to let the child turn on a faucet or a light switch, under supervision.

MONTHS 13 TO 18—The brain establishes accelerated and more complex associations, especially if the toddler experiments directly with objects. A rich environment will help the toddler make such associations, understand sequences, differentiate between objects and reason about them.

—SANDRA BLAKESLEE, April 1997

In Brain's Early Growth, Timetable May Be Crucial

FOR THE FIRST 28 MONTHS of her life, Simona Young languished in a Romanian orphanage. She lay in a crib alone for up to 20 hours a day, sucking nourishment from cold bottles propped over her tiny body. Unable to sit up by herself, she would push her torso up on thin arms and rock back and forth for hours, trying to soothe the aching void that had replaced her mother.

Now 6, she runs, talks and sings like other children her age. Since she was adopted by a Canadian family in 1991, she has been making steady progress, says her new mother, Jennifer Young. Yet problems remain. Simona still suffers temper tantrums and has trouble following spoken directions. She has difficulty sharing and taking turns with other children, and she will happily wander off with strangers who say kind words to her.

Psychologists at Simon Fraser University in Burnaby, British Columbia, are closely watching Simona's development, along with that of 44 other Romanian orphans who were adopted around the same time. Like Simona, 30 of the children experienced one or more years of profound deprivation in the overcrowded orphanage, where staff workers gave infants little or no personal attention. The other 15 were adopted within a month or two of their births.

The Canadian researchers are comparing the two groups of children to help answer an age-old question: Can love overcome a bad beginning?

Other scientists are asking similar questions, using the tools of modern cognitive neuroscience: Are there very early critical periods for emotional development? How does experience shape the brain's circuits? How changeable are those circuits later in life?

No one is saying there are quick fixes, that making nice to a baby between birth and 24 months will avert all later problems, said Dr. Carla

Shatz, a developmental biologist at the University of California at Berkeley and president of the Society for Neuroscience. But basic brain research is seeking answers that may ultimately help guide social policy, she said.

If there are critical periods for a child's emotional development, parents may be taught when and how to provide the kind of nurturing needed for healthy brain development. If the adult brain is amenable to change, maladaptive circuits formed in infancy or childhood may be alterable by psychotherapy or other methods.

Much is already known. Even the human fetus can hear sounds and has limited vision, Dr. Shatz said. "The nervous system isn't waiting for birth to flip a switch and get going," she said.

In the 1960s, Dr. David Hubel and Dr. Torsten Wiesel found that vision does not develop normally in cats if the eye and brain fail to make connections during a critical window of time in early life. Kittens that had one eye kept closed after birth did not develop the usual connections between that eye and the primary visual area of the brain. Once this period, lasting several weeks, had passed, none of the kittens could see out of the eye that had been closed, even though it was perfectly normal.

Hearing and language are also abilities that develop during critical periods, Dr. Shatz said. A Japanese baby can distinguish "r" from "l," but, absent the "l" sound in the Japanese language, loses this ability after age 3. After 10, most people cannot learn to speak a second language without an accent. Unless deaf children are exposed to some form of language before age 5, they behave as though they are retarded. And so-called "wild" children, raised without human contact, never learn to speak with fluency.

In recent years, the search for critical or sensitive windows of development has extended to other biological systems in the brain.

All animals, including humans, develop a control point in early infancy for how much of various stress hormones they will release in particular conditions, said Dr. Michael Meaney, a psychiatrist at McGill University in Montreal. Animals experiencing high stress levels in infancy develop a highly reactive system, he said, while animals raised in relative calm have quieter systems.

Dr. Myron Hofer, a psychiatrist at the New York State Psychiatric Institute in New York, has found numerous "hidden modulators" in the mother-infant relationship. For example, the licking of a mother rat influences the setting of her pup's heart rate, temperature, circadian rhythms, growth, immune system and other physiological states.

Other researchers are studying how a mother's touch literally helps shape her baby's brain. If baby rats are deprived of maternal licking when they are 7 to 14 days old, they develop fewer hormone receptors in their brains. Missing the needed stimulation in this critical period, they fail to grow normally, even when adequate amounts of growth hormone and insulin circulate in their tissues.

Human mothers provide similar modulators, Dr. Hofer said, through rocking, touching, holding, feeding and gazing at their babies. Some of these regulators are emotional, he said; thus, a baby knows when its mother is being cold or distant, despite her ministrations to physical needs. In the first six months of life, "the infant is laying down a mental representation of its relationship with its mother," Dr. Hofer said, adding, "These interactions regulate the infant's neural mechanisms for behavior and for feelings that are just beginning to develop."

If these early months of life are so important, what is actually happening inside the baby's brain? What kinds of changes are taking place?

At birth, according to Dr. Harry Chugani of the University of Michigan in Ann Arbor, a newborn brain has fewer synapses—connections between nerve cells—than an adult brain. (The same holds true for the complexity of dendrites, or branches.) But the number of synapses reaches adult levels by age 2 and continues to increase, far surpassing the adult level from ages 4 to 10, Dr. Chugani said. The density of synapses then begins to drop, returning to typically adult levels by age 16. These findings are based on direct anatomical measurements by Dr. Peter Huttenlocher of the University of Chicago, who measured the brains of children killed in car accidents, and on brain images from positron emission tomography (PET) scans that Dr. Chugani performed on infants for health reasons.

Concurrent with the explosion in the growth of synapses is a rapid pruning away of those that do not get used, Dr. Chugani said. There seems to be as much synaptic death as there is synaptic profusion.

The interplay between genes and experience in building a complex structure like the brain is to be expected, said Dr. Daniel Alkon, chief of the Neural Systems Laboratory at the National Institutes of Health in Bethesda,

Maryland. Human DNA does not contain enough information to specify how the brain finally gets wired.

Thus the newborn brain comes equipped with a set of genetically based rules for how learning takes place and is then literally shaped by experience, Dr. Alkon said. "This helps explain the power of childhood memories," he said. "Associations in early life help choose which synapses live or die."

Dr. Jeff Shrager, a neuroscientist at the University of Pittsburgh, says the infant's brain seems to organize itself under the influence of waves of so-called trophic factors—chemicals that promote the growth and interconnections of nerve cells. These factors are released so that different regions of the brain become connected sequentially, with one layer of tissue maturing before another and so on until the whole brain is mature. Such waves of chemical activity may help determine the timing of critical periods.

By the time the brain's production of trophic factors declines in later childhood, its basic architecture would be more or less formed, Dr. Shrager said. The process, since modulated by experience, would create human brains that were similar in their overall structure and interconnections but unique in terms of their fine connections.

The same trophic factor chemistry that makes young brains grow so dramatically may still be available in adulthood, particularly in the hippocampus, to help with adult learning and memory, Dr. Shatz said.

While that hope remains, there is a deeper question yet to be answered: Are there narrow windows in early infancy when emotional circuits are permanently established, or do emotional circuits form over many years so that early experiences are not so powerfully formative?

Much of the thinking is still speculative, Dr. Alkon said. "But we do know that a child learns trust and self-worth in the first two years," he said. "When a parent neglects a baby on a daily basis, the child is conditioned to expect isolation. A recipe for depression has been acquired from experience, handed down from one generation to another."

PET scans show that the frontal cortex becomes metabolically very active in infants aged 6 to 24 months, Dr. Chugani said, and again at puberty.

Thus it is possible that the frontal cortex—once thought to develop in later childhood—may be involved in early emotional and cognitive development, said Dr. Geraldine Dawson, a psychologist at the University of Washington in Seattle.

Dr. Dawson and others have found that the left frontal lobe is activated when a person feels happiness, joy or interest, while the right is associated with negative feelings. Infants of severely depressed mothers show reduced activity in the left frontal region, she said. Activity in the right is increased, which means the babies are vulnerable to negative emotions.

"Our hunch is that there may be a critical period for emotional development between ages 8 and 18 months," Dr. Dawson said. "This is when kids learn to regulate emotions. It is when attachment forms."

The insights into brain development gained from animal experiments might apply to humans, but in many cases repeating the experiments in children would be unethical. The Romanian orphans are of particular interest to brain researchers. The deprivations inflicted on them by the Romanian regime and, among the adopted children, the efforts of their new parents to nurture them back to normalcy, in effect constitute a unique experiment.

The children still in Romanian orphanages "look frighteningly like Harlow's monkeys," said Dr. Mary Carlson, a neuroscientist at Harvard University, referring to a well-known experiment of the 1950s in which baby monkeys were removed from their mothers a few hours after birth and reared without parental care. The infants developed abnormal behaviors. They often sat and stared for long periods, or would rock back and forth. Despite later efforts to rehabilitate them, the monkeys had disturbances in social behavior.

Many institutionalized Romanian orphans are below the third percentile in weight and height, Dr. Carlson said. Some show a profound failure to thrive, and at age 10 are the size of 3-year-olds, suggesting that the absence of early maternal interaction has had lasting effects.

But humans being more adaptable than monkeys, researchers are striving to reverse the effects of deprivation in the orphans. Elinor Ames, a psychologist at Simon Fraser University, notes that the older children, now 4½ to 10 years old, are catching up in language and physical development. But they are having trouble with social development. When they are in stressful situations, she said, they wrap their arms around themselves and rock for comfort.

The hope is that continuing good experiences will help these children grow into secure adults, Dr. Ames said, adding that she is optimistic.

But if it turns out that positive early experiences are crucial for healthy brain development and that deprivation leads to depression, anger and pathological behavior, what can society do to intervene? The question applies to babies being raised in violent neighborhoods by drug-addicted mothers, as well as to poor little rich kids whose parents are too busy to pay them attention.

Intervention experiments—in which disadvantaged children are taken to special day care centers from infancy to kindergarten five days a week and given a rich educational curriculum and loving environment—have worked, said Dr. Craig Ramey, a psychologist and educator at the University of Alabama in Birmingham.

Effects of the intervention did not begin to show up until the second year of life, he said, but at age 2 a matched group of children who were not given the intervention had IQ scores 15 points below those who were helped.

"The results are clear," Dr. Ramey said. "To make a difference, you have to intervene earlier" than Head Start. "We think we are affecting early mechanisms involved in language acquisition, in the depth and breadth of the language experience which lays a foundation for higher order thinking" later in life, he said.

Dr. Carlson, whose work with Romanian orphans has led her into advocacy for children's rights, said that she had been advised by people who worked for government child welfare agencies to play down the notion of critical periods and early brain development. "They say, 'If so much is determined by age 1 or 2, people will give up on children,'" she said. "But I think that if you believe in critical periods, you can find ways to take advantage of that plasticity."

Dr. Megan Gunnar, a professor of child development at the University of Minnesota in St. Paul, said the problem was not as difficult as people might think. "The one thing we have learned," she said, "is that children need to feel safe and protected. That alone leads to appropriate biological growth."

Dr. Hofer agreed. "If you grow up in battle-torn Yugoslavia, you may become impulsive, aggressive and you won't want to get close to anyone," he said. "You will be beautifully suited to fight a 500-year war in Europe."

—SANDRA BLAKESLEE, August 1995

Finding Elusive Factors That Help Wire Up Brain

THE HUMAN BRAIN is a natural wonder that boggles the mind. It is built of a hundred billion nerve cells sheathed and nourished by a trillion supporting cells. And projecting from every neuron is a delicate communications cable called an axon that splits and branches and splits again, like the crown of a great spreading oak, allowing that one nerve cell to cry out through synaptic conduits to a thousand others of its kind. All told, there are more synaptic connections encased within a single skull than there are stars flung across the universe.

Neuroscientists who study the developing brain have long sought to understand just how that extraordinary neuronal circuitry gets wired up in the first place. They have struggled to identify the signals responsible for axonal guidance—for steering the tender and tentative axon of a young nerve cell through the pandemonium of the growing brain and over to its appropriate target. Once the tip of the axon has found its correct resting spot, it can begin forming synaptic linkups with as many neurons in the neighborhood as it needs to contact to get its message across. And only with its axons in place can a brain begin to register the world around it: to see, to think, to feel, to know itself.

"It's been a kind of Holy Grail to identify the chemical signals involved in guiding the growing axons," said Dr. H. Robert Horvitz, a developmental neurobiologist of the Massachusetts Institute of Technology.

"It's been one of the classic problems in our field for the past 100 years," said Dr. Susan McConnell, a developmental neurobiologist at Stanford University.

Now that century-long effort is finally bearing fruit. In two papers in the journal *Cell*, Dr. Marc Tessier-Lavigne, Dr. Tito Serafini and their colleagues at the University of California at San Francisco have announced the discovery of the first two axonal guidance factors ever detected in vertebrates (animals, including humans, that have a backbone and a centrally organized nervous system). The factors are proteins generated by the budding nervous system when a human fetus is less than a month old and not yet the size of a pencil eraser. Diffusing outward from their point of origin, the factors help to guide the wiring of the spinal cord up to the appropriate centers of the brain where sensations like heat, cold and pain will eventually be registered.

The newly discovered factors also appear to control axonal movement in other major regions of the central nervous system, including the hindbrain and midbrain. And they are surely just two of scores if not hundreds of factors that jointly constitute the neuronal department of transportation—molecules that steer, instruct, heighten and subdue the traffic flow of millions upon millions of axons as they migrate restlessly across the fetal brain.

The researchers have christened the factors netrin-1 and netrin-2, after the Sanskrit word for "one who guides." Their work confirms predictions made in 1892, when the legendary neurobiologist Santiago Ramón y Cajal proposed that axons find their way to their destinations like bloodhounds on a leash, basically sniffing chemicals in their environment and proceeding accordingly. The netrins are like scent markers that tell the axons, "Yes, you're on the right path," or "Now it's time to veer sharply to the right and head north." For that reason they are called chemotropic factors: they are chemicals that persuade axons to turn and bend. They are also growth stimulants, prompting the axons to elongate as they migrate.

"This is certainly one of the most important and exciting discoveries in modern neuroscience," said Dr. Carla Shatz, a developmental neurobiologist at the University of California at Berkeley.

The new studies also evoke the image of the developing brain as a kind of Parisian parfumerie, where chemistry reigns and entices. And just as a particular perfume may attract one nose and repel another, so the netrins and other chemotropic factors in the brain seem to lure some axons to move forward, while causing the axonal extensions of other neurons to rear back in distaste and head in the opposite direction. In this way, a single factor can perform multiple guidance tasks in the developing nervous system.

Greatly delighting neuroscientists, the netrins turn out to be closely related to a factor already detected in a favorite staple organism of developmental biology, the roundworm, or nematode. Scientists studying the tiny translucent creature had demonstrated that a factor called unc-6 influences the migration of sensory axons and other cells as they circumambulate the worm's developing tubular form, although details of that control had remained elusive.

In the new work, Dr. Tessier-Lavigne and his co-workers show that the gene responsible for netrins is about 50 percent related to the gene that makes unc-6 in worms, a startling degree of genetic constancy given the 600 million years of evolution separating nematodes from mammals. The constancy is particularly impressive for a gene involved in threading the circuitry of the brain. If anything is supposed to distinguish an invertebrate worm from a vertebrate human it is the structure of the nervous system.

The latest observations offer the most vivid evidence yet that nature is basically lazy, even when designing brains, and that the difference between a worm's sensory apparatus and the exalted three-pound organ of humans is not one of quality, but quantity. In a review accompanying the new reports, Dr. Corey Goodman of the University of California at Berkeley talks about the human nervous system, particularly the spinal cord, as "the worm within us."

Beyond justifying the deeply held convictions of worm workers that their organisms are reasonable facsimiles of people, at least in the laboratory, the similarity between the two classes of factors means it will be much easier to delve deeper into how the guidance molecules of the brain operate. "We can go back and forth from the mammalian work to the nematode studies, and the field will progress very rapidly now," said Dr. Goodman in a telephone interview.

That ease, in turn, may eventually yield new approaches to treating neurodegenerative diseases, spinal paralysis and other disorders of the central nervous system. Scientists might attempt to use a netrin-like drug to foster the regeneration of severed nerves, for example. But though the idea is encouraging enough that Dr. Tessier-Lavigne's research is supported in part by the Paralyzed Veterans of America Spinal Cord Research Foundation, the scientists emphasize that they have just begun to decode the brain's surpassingly dense wiring diagram.

As much as Ramón y Cajal has been revered by contemporary neuroscientists, his proposal that axons find their way with the aid of chemotropic factors surrounding them has not always been accepted. Some scientists believed axons arrived on the scene pre-labeled, with their eventual destinations hard-wired into their specific genetic codes. Others believed that anarchy reigned in the developing brain, and that axons hooked up with targets more or less randomly, and then those connections that worked were locked into place by subsequent use. However, both extremes had their conceptual problems, and the compromise notion of brain wiring by chemotropism persisted. In this scheme, axons have a vague pre-programmed idea of what chemicals in their environment they should respond to, but they need the help of factors like netrins to reach their destination.

Beginning in the mid-1980s, scientists began gathering evidence that chemical cues controlled the migration of axons during development. Dr. Tessier-Lavigne, Dr. Thomas M. Jessell of Columbia University and others worked on an embryonic structure called the floor plate, one of the earliest parts of the nervous system to mature and a crucial player in organizing the rest of the spinal cord and brain. In experiments, they found that the cells of the floor plate had a magnetic effect on neurons—put the floor plate together with neurons in a bit of gel, and the axons of embryonic nerve cells stretched toward the floor plate like the arms of movie zombies.

However, isolating the product within the floor plate cells that had this effect proved enormously difficult. To find and purify the netrin proteins, the Tessier-Lavigne laboratory had to pulverize about 25,000 embryonic chick brains over a period of several months. However, the labor paid off. The scientists determined that the netrins alone had the same attractive effect on axons as had the entire floor plate structure.

The researchers propose a rough model of how netrins influence the behavior of axons extending from so-called spinal commissural neurons—nerve cells that convey to the brain information about pain and temperature. By the third or fourth week of human embryogenesis, the floor plate has matured into a structure located in the middle of the nervous system, toward the front of the primitive spinal column. Once it has matured, the floor plate begins secreting netrin proteins into the surrounding fluids. Heeding the signal, rudimentary neurons located at the back of the budding spinal column begin projecting their axons. At the tip of the axon is the bloodhound, a fidgety cell-like structure called the growth cone, which can sense chemicals in the environment, presumably through receptor proteins studding its membrane sheath.

While the neuron stays put, the growth cone pulls the axon around the circumference of the spinal column toward the netrin cologne wafting from the front. The axon crosses the midline ridge of the spinal cord, and then, again at the nudging of netrins, makes a 90-degree turn and heads up into the brain, where the growth cone can connect with cerebral centers designed to accept sensory information. Once they reach their target position, the growth cones mature into so-called synaptic terminals, the sites of communication that allow signals from one neuron to jump and spark to another neuron and pass a thought or impulse along.

In taking their circumferential course around the spinal column, the axons from neurons in the right side of the spine connect to centers on the left hemisphere of the brain, while neurons on the left connect to the right. For this reason, a person who has a stroke in the right side of the brain ends up with paralysis on the left side of the body.

"We still don't understand why the nervous system is organized in this fashion," said Dr. Tessier-Lavigne.

The netrins appear to be important mostly for helping the first generation of spinal axons to make their way. Once these pioneers have mapped out the appropriate course, axons from other neurons follow suit, wrapping themselves beside and around the original cable.

And even when netrins have steered the axons toward their initial connections, the wiring of the brain has only just begun. Often many axons end up at the same resting spot, at which point they must compete for the privilege of connecting to neurons in the neighborhood. Those axons that fail to win a linkage site within hours or days retract and die.

What is more, the brain is refined and re-refined through use. As signals from the outside world begin impinging on the brain through the eyes, the ears, the nose, the spine, the synaptic connections between neurons become stronger and evermore elaborate. The refinement of the human nervous system continues throughout life, which is why people can master new skills or think revolutionary thoughts into old age. So while the netrins and other chemotropic factors have laid out the basic circuitry of the central nervous system by the second month of fetal development, it is living that infuses it with life.

—NATALIE ANGIER, August 1994

Evolution of Tabby Cat Mapped in Brain Study

Felis silvestris tartessia, Spanish wildcat

Michael Rothman

BY COMPARING the brains of Spanish wildcats and American domestic tabby cats, a researcher in Tennessee has discovered a biological mechanism that may explain how members of a species may adapt their brains to undergo rapid evolutionary change.

The finding is that wildcats and domestic cats develop the same number of brain cells as fetuses, but in each species different sets of neurons are killed off just before birth. The result is presumably that each cat's brain is better adapted to fit its environment.

Thus the wildcat retains nerve cells that mediate excellent color vision and enable it to hunt in the bright Iberian sunlight, said the researcher, Dr. Robert Williams, an assistant professor of anatomy and neurobiology at the University of Tennessee College of Medicine in Memphis.

But the domestic cat, which is nocturnal, discards most neurons for color vision and instead nurtures cells that sense motion and objects in dim light. Other important brain areas are also sculptured differently in the two cats, Dr. Williams said.

The immense evolutionary advantage of adapting to different environments by killing off selected brain cells before birth, Dr. Williams said, is that the animal retains the ability to re-evolve traits rapidly should the world change abruptly. Thus the domestic cat has the latent capacity to redevelop rich color vision should it ever need to switch to daytime hunting, he said.

Compared with closely related wildcats, the domestic cat has lost 30 to 50 percent of its brain cells in adapting to the lap of luxury, Dr. Williams noted, although this does not mean pussycats are stupider than wildcats. Each animal has an intelligence honed for making a living in its chosen niche, he said.

Dr. Williams's work, described as the first experimental study linking cell death with brain evolution, appears in *The Journal of Neuroscience.* His co-authors are Dr. Carmen Cavada and Dr. Fernando Reinoso-Suárez at the University of Madrid.

Dr. Leo Chalupa, a professor of neuroscience at the University of California at Davis and Dr. Williams's thesis adviser in the mid-1980s, said his former student "really deserves a lot of credit for being innovative and clever." He added, "Very few people in neuroscience take an evolutionary perspective, and this is a very important finding."

Dr. Harvey Karten, an expert on evolutionary biology at the University of California at San Diego, was more cautious. "To say that selective fetal cell death is a mechanism for rapid evolution is an interesting idea," he said. Dr. Williams "demonstrates the loss of color vision in the domestic cat in just 20,000 years, which is extremely rapid," Dr. Karten said. However, he said, "to say cell death is the mechanism through which evolution expresses itself goes beyond current data."

Dr. Murray Sherman, an expert on the cat visual system at the State University of New York at Stony Brook, said it was not unusual for closely related species and even subgroups of the same species to have very different visual systems. "The real surprise is the speed at which the change occurred in the cat," he said. "The idea that a part of the brain can evolve so differently in so short a time makes one wonder about different human races. Are we different in subtle ways?"

In mammals, up to half of all fetal brain cells are killed off before birth, Dr. Williams said in a telephone interview. It happens in every brain structure examined. For example, an adult human has 1.2 million to 1.5 million ganglion cells in each eye, while a third-trimester human fetus has 2.5 million such cells in each eye.

But cats exhibit an even greater level of cell death in the developing eye, Dr. Williams said. An adult has 150,000 ganglion cells while an unborn kitten has 900,000 cells per eye. "The domestic cat loses four out of five ganglion cells produced before birth," he said. "Why?"

The fossil record shows that the domestic cat shrank in size over the last 20,000 years, Dr. Williams noted. It is half as large as the wildcats from which it descended, so perhaps its brain shrank as well.

But how? A brain can shrink by making fewer cells, Dr. Williams said, or by packing the same number of cells more densely or by killing off more cells before birth.

It occurred to him that the answer might emerge in comparing the brains of wildcats and modern domestic cats. By a fluke of nature, a wildcat species—*Felis silvestris tartessia*—has survived unchanged for the past 20,000 years in the mountains of Spain, Dr. Williams said. The Gulf Stream protected the Iberian peninsula from the ravages of ice ages and rapid warming so that the wildcats' environment has remained amazingly stable, he said. As living fossils of the Pleistocene, the tabby-like cats successfully evaded human contact while their cousins in Europe and Africa evolved into domestic species.

Several years ago, game wardens captured two of the solitary wildcats, Dr. Williams said. Although the animals usually weigh 14 pounds, these were skinny and so badly injured that they could not be released into the wild. After attempts to breed them produced no kittens, the cats were sacrificed for scientific study.

Dr. Williams compared them with a domestic male tabby cat that weighed nearly 20 pounds. While most domestic cats weigh about seven pounds, a larger cat was used to rule out the effect of body weight on brain size.

Surprisingly, the brain of the fat tabby weighed 28 grams, but that of the male wildcat, thin as it was, weighed 37 grams, or nearly a third more. Moreover, the domestic cat's skull was twice as thick as the wildcat's.

Though the retinas of the two cats were the same size, that of the wildcat packed 100,000 cones—cells that specialize in color vision—onto each square millimeter, compared with the mere 35,000 cones of the domestic cat. The wildcat had twice as many ganglion cells that connect the eye to the brain and a third more cells in the brain's first visual relay station behind the optic nerve.

But cells that facilitate black-and-white vision, motion detection, object recognition and perception in dim light remain the same in both cat species, Dr. Williams said. The part of the brain that helps a cat catch a mouse in the dark is intact.

Dr. Williams concludes that the domestic cat lost color vision over the past 20,000 years, which in evolutionary terms is extremely rapid. It is not likely that the Spanish wildcat gained color vision in that period of time, he said, since the trait is extremely complex. A more likely explanation, in his view, is that domestic cats pruned away most cells for color vision before birth because they had less need for them.

The rest of the domestic cat brain has undergone similar changes, Dr. Williams said.

Unbeknownst to the gamekeepers, Dr. Williams said, the female wildcat was pregnant with a single kitten. Upon examination, its brain contained a similar number of neurons as a developing domestic kitten of the same gestational age.

Thus programmed cell death in the developing brain may explain why animals that live in stable environments undergo little evolutionary change while animals under environmental pressure can change extremely rapidly, gaining or losing traits with astonishing speed. Such changes do not occur in one generation, Dr. Williams said, but take thousands of years.

Domestic cats probably got smaller in response to climate shifts and became nocturnal in response to hanging around villages filled with hostile humans, Dr. Williams suggested. With the exception of ancient Egyptians and a minority of modern humans, he said, people have thrown rocks at, hung or tortured cats more than they have cuddled them. For a cat in human proximity, it makes sense to hunt or steal food at night when people are asleep.

Domestication may have altered the size of the domestic cat's amygdala, a brain center that controls aggression and docility, Dr. Williams said. He plans in further studies to look at the amygdala and other brain regions of the two cats.

But why the domestic cat's skull is thicker remains a mystery. Although there is a folk saying that "the thicker the skull, the dumber the animal," there is no evidence domestic cats are stupider than their wild brethren, said Dr. John Gittleman, a zoology professor at the University of Tennessee and expert on carnivore evolution. There are many kinds of intelligence, he said, and the number of brain cells is less important than the way in which they are connected.

Dr. Gittleman noted that domestic cats have assumed many infantile traits, like snubby faces and "popcorn behavior," meaning they jump around without warning. Perhaps these traits, too, are the result of rapid evolution. It makes one wonder if there is a cell death program for cuteness, he said.

—SANDRA BLAKESLEE, January 1993

Rare Gene Flaw May Help Explain Shaping of Brain in Womb

AS YOUNG BOYS, the patients often suffer from two outstanding but seemingly unrelated symptoms: They have no sense of smell, and their genitals are abnormally tiny. Scientists had long wondered why such disparate problems show up in a single genetic disease, and now they think they have the answer.

Their insights could result not only in better diagnosis and treatment of the rare disorder, called Kallmann's syndrome, but also in helping illuminate the fundamental mystery of how the brain shapes itself during the growth of a fetus.

Reporting in *The New England Journal of Medicine,* Dr. Andrea Ballabio of Baylor College of Medicine in Houston and his colleagues presented proof that they have isolated the gene which when defective causes the syndrome, and they have a compelling theory of what the gene normally does in the body.

They propose that the healthy gene directs the production of one of the air traffic controllers of the developing brain, a molecule that steers restless young nerve cells toward their proper destination. Many proteins are thought to orchestrate the frenetic flux of neurons across the flowering brain, nudging some toward the visual regions, others in the direction of, say, the limbic system, where emotions arise. In this case, the factor specifically ushers one population of cells toward those neural neighborhoods in charge of smelling and of releasing sex hormones.

But Kallmann's patients lack this critical escort molecule, and so the neurons the factor is meant to guide get stuck uselessly where they originated, deep in the forefront of the embryonic brain. As a result, patients are born without any olfactory nerve fibers and without cells in the hypothalamus needed to pump out gonadotropin-releasing hormone, a modulator of sexual development.

The new discovery, which builds upon previous work by Dr. Ballabio and others, is the first nerve-cell migration factor to be studied as the cause of a human disease.

"This is very important work, and it is of great interest to many in the field," said Dr. Aaron Moscone, a professor of developmental neurobiology at the University of Chicago. "Right now there are many labs trying to understand how neurons recognize their position and associate with the right partners. The vocabulary of the brain's recognition molecules is turning out to be quite large and quite complicated. Kallmann's syndrome is one way to understand that language."

Researchers hope that by understanding the disease, they will have a grasp on other disorders thought to be caused by a flaw in nerve cell journeys, including some types of epilepsy and a bizarre and lethal syndrome called lissencephaly, in which babies are born without any folds on the surface of their brains.

"We think this is very exciting," said Dr. Ballabio. "Kallmann's is the first example of a neuronal migration defect in man or any other vertebrate in which the gene has been identified. This could be of real use to both patients and to basic scientists."

For families in which Kallmann's syndrome is common, the isolation of the gene offers a way of identifying the disorder as early as possible, to begin a treatment like the use of hormone replacement therapy. Kallmann's patients given synthetic testosterone in childhood can end up with normal genitals, although results vary.

The newly isolated gene could also be used for prenatal diagnosis, but Dr. Ballabio said he doubted that many people would choose to abort a fetus if they found it carried simple Kallmann's syndrome. "The disease is not terrible, and it can be treated," he said. "Kallmann's patients live a perfectly normal lifespan."

A defect in the Kallmann's gene is often accompanied by other flaws in the surrounding regions of the chromosome, however, and such multiply afflicted babies can end up severely handicapped or retarded. Dr. Ballabio proposes that the new genetic marker may be useful for detecting the more extreme forms of mutations early in pregnancy.

Kallmann's syndrome is quite rare, afflicting fewer than one in 10,000 newborns. A vast majority of Kallmann's patients are men, because the gene behind the disease is on the X chromosome. Women, with their double X chromosomes, harbor two copies of the crucial gene, and thus are protected against the effects of a mutation in one of those genes. By contrast, men, with their one X and one shorter Y chromosome, have only a single version of the gene, and if that copy is perturbed, they end up with the disorder.

Women can act as carriers of the disorder, however, and pass it onto their sons. The most common symptoms of the disease are lack of smell and genital aberrations, including sterility, but some patients also have misshapen kidneys and display a trait called mirror movements.

"If you ask the patients to twist one hand, they'll twist the other hand as well," said Dr. Ballabio. "They won't do it if they're paying careful attention, but they will spontaneously." Somehow, in these patients the usual inhibitory mechanism that prevents one side of the brain from mimicking the other side's activities is disturbed, and so both lobes tell their respective extremities to move simultaneously.

The researchers, along with scientists in Italy, hunted down the gene by studying 77 families in which one or more men suffered from Kallmann's. Using genetic and molecular techniques, they sought revealing defects on the X chromosome that were confined to those with the disease. Eventually, they came upon a gene that bore striking similarity to other genes known to produce so-called axonal pathfinding molecules, most of which have been studied in fruit flies and mice.

The factors guide developing neurons along particular pathways in the brain and allow the neurons to make their proper connections with other cells once settled. Scientists have scant idea how the pathfinding factors do their job and where exactly in the brain they operate. Some probably assume a commando position while based on the surface of connective tissue, directing oncoming neurons from a seated position, while others seem to circulate freely through the fluids of the brain.

The resemblance between the Kallmann's gene and other migration factors fitted current models of the syndrome, and the researchers confirmed their discovery by studying two young brothers. One had been born with a clearly stunted penis, while the testicles of the second began to retract by the age of four months. Both proved to have a modest mutation right in the middle of the gene isolated by the Baylor researchers and both responded well to hormone treatment, their genitals quickly assuming normal proportions.

—NATALIE ANGIER, July 1992

A Brain Cell Surprise: Genes Don't Set Function

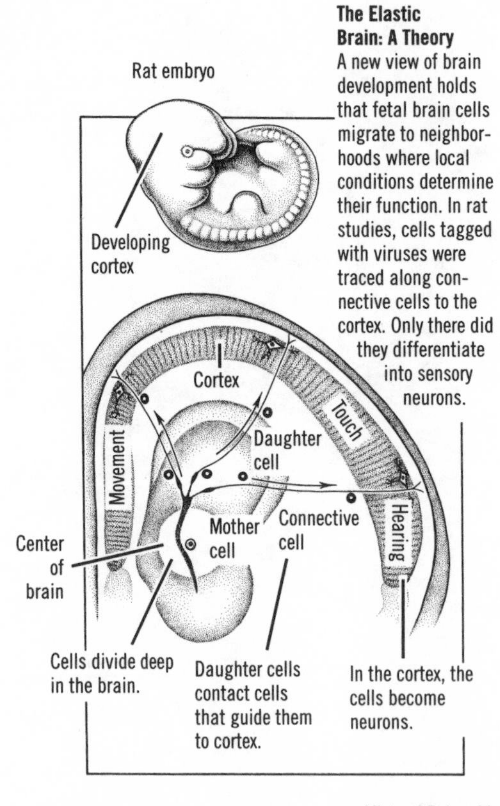

The Elastic Brain: A Theory
A new view of brain development holds that fetal brain cells migrate to neighborhoods where local conditions determine their function. In rat studies, cells tagged with viruses were traced along connective cells to the cortex. Only there did they differentiate into sensory neurons.

Michael Reingold

YOUNG NERVE CELLS can wander far greater distances across the growing brain than anyone previously imagined, two scientists have found. And once the neurons have settled into a particular neighborhood, they learn what they are meant to do from signals that surround them, rather than from an innate genetic program, the researchers said.

The new research strongly supports the idea that the cerebral cortex, the most advanced component of the mammalian brain, is extremely supple and responsive to its environment, instead of being fixed early on by a blueprint in the genes, as some scientists had believed.

By the latest theory, a neuron that ends up in the visual region of the brain becomes a visual neuron not because its genes so instructed it, but because the cell is exposed to the appropriate signals from nearby cells and from the optic nerve. If that same fledgling neuron had landed in the auditory region of the brain, the scientists said, it would have matured into a neuron fit for hearing.

In the last several years many biologists have come to favor the idea of neuronal plasticity, but the latest report offers the most detailed and powerful confirmation yet of the theory.

Dr. Constance L. Cepko and Dr. Christopher Walsh of Harvard Medical School in Boston demonstrated that otherwise identical neurons descended from the same mother nerve cell will, during brain development, journey to wildly different regions of the cortex and assume very different careers once settled.

The report appears in the journal *Science.*

"We're normally thought of as nihilists, and so we didn't think anything could surprise us," said Dr. Walsh, who is also a neurologist at Massachusetts General Hospital, referring to his philosophy that many prior conceptions turn out to be wrong.

"But even we were surprised by how far these guys would travel" from the point where the dividing parent neuron had spawned them, he said.

The researchers do not yet know what determines when a young neuron will cease migrating across the cortex and join a particular spot of the brain, but they believe external cues like a tempting vacancy are probably far more important than any genetic instructions the neurons carry with them.

"The mother cells do not impart specific information to their daughters about what to become," Dr. Cepko said. "They do not say to one cell, you go to the visual cortex and to another, you go to the motor cortex."

The researchers tracked the patterns of cell migration by using a technically daunting procedure, first injecting telltale molecular tags into the developing brains of fetal rats, and later screening brain cells to determine which neurons had landed where.

"I think this is one of the coolest pieces of work I've ever read," said Dr. Carla J. Shatz, a professor of neurobiology at the University of California at Berkeley. "Doing this kind of experiment is like climbing Mount Everest. It's not the sort of thing many people are willing or able to undertake. They did a beautiful job."

Dr. Dennis D. M. O'Leary, a neurobiologist at the Salk Institute in San Diego, California, who has also shown neural adaptability by transplanting brain tissue in fetal rodents, said, "It was a technical tour de force and a very seminal study."

Neurobiologists have been arguing for decades over whether embryonic neurons are blank slates or pre-fabricated units destined for a particular fate. Some studies suggested that during development, related neurons move in lockstep along set paths to reach one region of the cortex or another, as though under genetic command, but until now scientists could not follow cell migrations precisely enough to understand unequivocally what they were seeing.

Conversely, other experiments seemed to indicate that neurons are enormously flexible. Researchers demonstrated that if they moved a tiny piece of tissue from a fetal rat's visual cortex—the region of the brain that controls sight—and grafted it onto the rat's somatosensory cortex, which rules over body sensations, the transplant would take on all the features of its new location, becoming a swatch of the somatosensory cortex.

In an effort to settle the debate, Dr. Walsh and Dr. Cepko devised an exceedingly sophisticated technique. They knew that the brain develops as a tube, with precursor cells within the tube dividing to give rise to offspring neurons, which then migrate outward in all directions along underlying connective fibers. "It's designed like spokes on a wheel," Dr. Walsh said.

But the precursor cells, or mother neurons, remain within the tube, providing a steady pool of daughter cells to flesh out the brain. The Boston researchers decided to try marking different mother cells with distinctive tags, which would then be passed on to their children and could be followed wherever the daughter cells should roam.

The scientists began with 100 different strains of viruses, intended to serve as easily identified fingerprints. Anesthetizing a pregnant rat at the 15th day of gestation, while the embryonic brains within her were at an early stage of development, the researchers injected the viruses deep into each pup's neural tube, where they infected some of the precursor neurons.

Because so many different types of viruses had been introduced into the growing brains, the odds greatly favored each neuron's being infected with a unique viral strain that then became its unmistakable molecular signature.

The operation had no apparent effect on the fetal rats, and the animals developed and were born normally.

A week or so later, the scientists removed the rodents' brains and then isolated cortical tissue. Using a technique called polymerase chain reaction to discriminate between viral markers in individual neurons, the biologists saw astonishing patterns of cell distribution. Daughter cells from many different mother neurons were sitting side by side in the same cortical neighborhoods. Equally impressive, siblings of the same mother cell had meandered to all points on the globe of the cortex, taking up residence in many different functional regions.

Judging by the relatively casual and widespread distribution of neurons, the researchers concluded that the cells were not likely to be obeying strict maternal advice inscribed into their DNA. Instead, the scientists propose that when the neurons arrive at their final location, they are hailed by other cell components—long feathery fibers called axons—which hook up to the neurons and instruct them on their assignment.

"The neurons are greeted by axons coming from other parts of the brain, and the axons tell them precisely what to do," Dr. Walsh said. "One set of axons carries visual information, another set carries information about sensation."

In the visual cortex, for example, axons may extend backward into the brain from the optic nerve, relaying visual pulses inward that in turn influence the local neurons.

"Development can either proceed by the British plan, where one's fate is determined by one's ancestors, or by the American plan, where one's fate is determined by one's neighborhood," said Dr. Walsh, using an analogy first suggested by the British geneticist Sydney Brenner. "The cortex turns out to be on the American plan in the extreme."

The researchers admit that their work engenders more questions than it resolves, and that in a sense it is a negative result: The experiment says what the neurons of the cortex are not. They are not pre-fixed units outfitted with a complete code of how to behave, and they almost surely do not provide the organizing principle of the mammalian brain. Other researchers are struggling to find out what is then the master plan behind the body's gray matter. If the cortical neurons are clued into their assignment by axons that are already in place, what signals told the axons where to go and what to do? What tells an axon that it is meant to help the brain see, or smell or hear?

Many researchers suspect that brain development proceeds in a great and exquisitely complex feedback loop, partly determined by the oldest regions of the brain, like the hindbrain that controls breathing, and partly by signals coursing in from distant regions of the flowering nervous system.

They also believe that the thalamus, a region deep in the midbrain where many sensory signals converge, develops before the cerebral cortex and then helps shape the cortical layers that grow up around it. But the basis of the architecture of the higher brain remains a subject of intense investigation.

Whatever the lingering mysteries, the new results do suggest why the mammalian brain was able to evolve so quickly and to such an impressive degree of complexity. Because neurons of the cortex seem to be remarkably malleable, able to adjust their performance depending on the input they receive, they can swiftly respond to changes in the rest of the animal's body without having to bother going through cumbersome genetic alterations themselves.

For example, the paw of a primitive rodent can evolve into the wing of a bat, but the neurons that interpret sensory signals from the limb need not be re-programmed to manage the change in body plan. That degree of elasticity in the brain may have had all sorts of spinoffs, including, eventually, the evolution of consciousness.

—NATALIE ANGIER, January 1992

6

DAILY RHYTHMS AND SLEEP

The brain can create art and math and flights of fancy but it also has functions of a humbler sort, such as keeping time. The clock in the brain tracks the all-important cycle of day and night. The signals from the clock govern the daily rhythms that are evident in functions like hormone production and the alternation of sleep and wakefulness.

In humans the clock is located in a barely visible cluster of cells called the suprachiasmatic nucleus. The basic rhythmicity is generated deep within each nerve cell, in interactions between the cell's nucleus and surroundings, and the cells probably contact each other to determine a consensus rhythm.

From studies of mice, the basic rhythm generated by the clock is known to be entrained, or kept in pace with the 24-hour cycle, by the light the animal perceives. The body's clock, an ancient mechanism found even in bacteria, works for the most part in silent efficiency. But it can become seriously confused by events for which evolution never prepared it, such as switching from day to night work, or flying through several time zones in a single day. The clock may take several days to reset itself to the new time phase of the traveler's destination, and during this period its owner may suffer the unpleasantness of jet lag.

The nature of the clock's influence over the body's rhythms is a matter of active study, as is evident in the chapters that follow.

Modern Life Suppresses an Ancient Body Rhythm

AS THE VERNAL EQUINOX advances, and the sun lingers in the sky a bit longer each day, and the buds poke forth like babies' fists from every barren twig, even urbanites may feel the pagan craving to revel in seasonal rhythms.

After all, the lengthening of the day and the warming of the air exert a tremendous influence on virtually every other life form, inspiring migrations, ending hibernations, inciting growth and exciting lust. Surely humans, too, must be prey to the power of the seasons, the return of light and the chastening of night. Surely people's innate circadian clocks must react to the return of spring, resetting themselves to keep pace with the extra daytime hours.

As it turns out, human biological clocks do change but only in about half of all people—the half who are women. In men, however, the songs of the seasons apparently hit a deaf ear. The contemporary industrialized world, which blazes with artificial illumination, has suppressed men's ability to react to changes in day length.

Women and men may sleep the same number of hours each night, they may spend the same amount of time bathed in a corporate fluorescent glow each day, but in women, at least one essential keeper of internal circadian rhythms ignores false idols of light and instead heeds only Helios, Ra, Mithras—the sun. When the sun rises late and sets with sorry haste, the amount of a key circadian hormone, melatonin, that is secreted in the female brain at night increases. Come summer, nocturnal melatonin release falls off.

The consequences of that seasonal hormonal shift remain unknown for humans, but in other species, annual changes in melatonin secretion serve as the principal signal orchestrating many of the behaviors that count, including a willingness to fly thousands of miles to one's summering grounds and the desire to breed.

Among modern men, by contrast, though they retain all the machinery to react to seasonal change, the release of melatonin at the winter solstice is identical to that secreted during a midsummer night's dream.

"Men seem to be more sensitive to artificial lights than women are," said Dr. Thomas A. Wehr of the clinical psychobiology branch of the National Institutes of Mental Health, who made the discovery of gender differences in circadian rhythms. Looked at another way, he said, grinning slyly, "When it comes to seasonal change, men just don't get it."

The surprising new finding is part of a larger study that Dr. Wehr and his colleagues are carrying out in the relatively unmined field of photoperiodicity in humans—the impact of day length on hormonal fluxes, sleep patterns and behavior. They are seeking to measure key indices of seasonal rhythms in humans and to see when, why and how those measures might change over the course of the year. Their work could explain why women suffer disproportionately from seasonal affective disorder, or SAD, a type of depression that strikes most often in winter.

In addition to the study of seasonal shifts, the researchers are attempting to tease apart the details of the body's circadian clock, which operates on a 24-hour schedule and tells a person when to sleep, eat and be out and about. Through elaborate and demanding studies that require volunteers to be hooked up, prodded, bled and sampled like astronauts for weeks at a time, the scientists are attempting to determine what the human circadian clock may have behaved like in prehistoric times, before the advent of bright lights, big cities and all-night cybersex.

"It's a kind of archeology, or human paleobiology," said Dr. Wehr. "We're looking at what human hormonal, sleep and temperature patterns might have been like in a prehistorical period when there was very little artificial light around."

Some of the scientists' results are preliminary, and those describing gender differences in seasonal release of melatonin have yet to be published. Nevertheless, the work suggests that women and men live in slightly different nightly realms, and that women can add to the lunar timetable on which menstrual cycling is roughly based a gentle adherence as well to the calendar of the sun. However, Dr. Wehr points out that there may be male cadences as well, perhaps ones operating on a shorter timetable.

The new studies are part of a larger explosion of interest in biological clocks. Several weeks ago, scientists announced the discovery of a gene in plants that controls such circadian-based rhythms as the morning unfurling of leaves and the timing of photosynthesis. Writing in *American Scientist,* Dr. Joseph S. Takahashi, a professor of neurobiology and physiology at Northwestern University, describes efforts to fish out the genes responsible for timekeeping in animals. One has been found in fruit flies, called the period gene, which assures that newly mature flies will emerge from their pupal cases in the morning, when the sun can quickly dry their wings. Another gene identified in the fungus *Neurospora* controls growth spurts. Dr. Takahashi and his co-workers are closing in on a gene called clock, found on both mouse and human chromosomes, that, when mutated, causes the body's clock to think the world works on a 25-hour day.

When Dr. Wehr began studying photoperiodicity in humans, he was astonished to learn how little work had been done in the area. Animals, yes—their seasonal variations had been charted in detail. But studying the same thing in humans is hard, long-term work. Besides, many scientists assumed that people, being non-seasonal, year-round breeders and non-migratory, were relatively immune to the effects of changing day length.

To explore the accuracy of such assumptions, Dr. Wehr and his colleagues attempted to recapitulate prehistoric sleep conditions in a post-historic population, among 15 young men living in the Washington area. They began by exploring what happens when the men switched from the conventional day length, or photoperiod, of 16 hours, which the average working adult adheres to year-round with the help of lamps and coffee, to a shortened photoperiod of only 10 hours, a schedule that approximates what prehistoric people in the middle latitudes would have experienced in the dead of winter.

Every night for four weeks the men came to the laboratory, where they spent 14 hours in windowless dark rooms, relaxing and sleeping as much as possible. Various hormone levels, temperature, brain waves and other functions were measured at regular intervals throughout the night. Later, similar measurements were made when the men came into the clinic to sleep for the more traditional seven to eight hours a night.

The researchers discovered a number of intriguing things about how ancestral humans may have spent their dark winter nights. For one thing, as the study volunteers adjusted to their artificial circumstances, their sleep patterns relaxed into distinct phases. The men slept only about an hour more than normal, but the slumber was spread over about a 12-hour period. They slept for about four to five hours early on, and another four to five hours or so toward morning, the two sleep bouts separated by several hours of quiet, distinctly non-anxious wakefulness in the middle of the night. The early evening sleep was primarily deep, slow-wave sleep and the morning episode consisted largely of REM, or rapid-eye-movement, sleep characterized by vivid dreams. The wakeful period, brain wave measurements indicated, resembled a state of meditation.

"This is a state not terribly familiar to modern sleepers," Dr. Wehr said. "Perhaps what those who meditate today are seeking is a state that our ancestors would have considered their birthright, a nightly occurrence." Dr. Wehr pointed out that many mammals, while they are secure in their dens, also sleep in bouts separated by stretches of quiet wakefulness.

During the enforced long nights, the levels of the hormone prolactin, which ordinarily double during sleep over daytime concentrations, switched into high gear shortly after the men entered their chambers, and remained elevated during the entire 14 hours. That perpetual prolactin surge could explain the overall peacefulness the men experienced. Prolactin is a compound that helps keep an animal resting; when birds brood their eggs, they stay still thanks to prolactin. The men's melatonin release also kept pace with the contrived night. The pineal gland of the brain began secreting melatonin at the onset of darkness and it continued pumping the hormone out at fairly high levels until morning. The output of growth hormone, a chemical associated with bodily construction and repair, also remained high throughout the night.

When the men returned to the normal schedule of an eight-hour night, a host of fairly dramatic changes occurred. Melatonin secretion dropped off. The amount of prolactin released stayed the same, but it was compressed into a much shorter time frame. The peak of maximum growth hormone output during the night doubled over what it had been on the winter schedule. Body temperature stayed slightly warmer than it had been during the long nights.

These and other alterations in physiology persuaded the scientists that humans, contrary to previous notions, have retained their innate ability to respond strongly to changes in the length of day and night. How those seasonal adjustments might have affected the behavior of early man is unknown. Nor is it known whether people living in the tropics, where the length of day and night is constant throughout the year, would respond to changes in light conditions, or whether cues like weather and temperature changes might have the same effect on melatonin secretion and the like.

More recently, in a paper in the *American Journal of Physiology*, the Wehr team showed that these seasonal changes were suppressed by the impact of artificial lighting and the preferred Western pace. They took men who had been going about their daily lives on the standard pan-seasonal schedule of 16 waking, illuminated hours and put them in a dimly lit room for 24 hours, where they were kept awake and again sampled regularly for things like melatonin secretion. Because the body's clock operates by a memory that lags for a day or two, the samples reflected the men's hormonal indicators of day and night from the time preceding their foray into the laboratory. Measurements were taken for the same man once in winter and again in summer. And while the scientists' previous studies showed that men can change their biological clocks from one season to the next when given the opportunity, these men, plucked from the real world for a day, showed no seasonal variation at all in melatonin secretion and other parameters.

The scientists found a gender difference in the seasonal response, not by looking for it, but as an incidental outcome of another study, an attempt to find the root cause of seasonal affective disorder. They wondered if sufferers of SAD might have seasonal variations in their melatonin secretions. Because most patients with the disorder are women, the researchers used healthy women as their control population. Subjecting SAD patients to the all-night sampling vigil they had put the men through earlier, the scientists found there was a significant difference in winter and summer pacemaking in the women with the affective disorder.

But to their astonishment, the investigators saw the same seasonal discrepancy in the healthy women as well. The winter patterns of melatonin secretion were long and more sustained; the summer pattern, shorter and with a lower maximum crest. All the women had been living according to the standard 16-hour up-and-ready schedule adhered to by their male counterparts. The women's internal clocks, however—the part of the brain that responds to light and dark—were somehow keeping track of the seasons and adjusting melatonin output accordingly.

Why women might be more resolutely photoperiodic than men remains a mystery. It could be a remnant of a time when humans, like many other mammals, tended to reproduce seasonally. In recent years, several scientists have found indications that humans still have semiseasonal breeding patterns, with peaks of fertility during spring and fall.

Beyond the gender difference, the work underscores how radically the change in human sleep patterns may be affecting biology and even behavior. By compressing all nocturnal biochemistry and all sleep patterns into an eight-hour period pretty much year-round, Dr. Wehr said, "we essentially live in an endless summer, from the day we are born until the day we die." The consequences of that compression have yet to be charted.

Nor does Dr. Wehr suggest we should return to paleolithic ways of long, languid nights. "We like being awake and doing things, and we don't want to go to bed early," Dr. Wehr said. "We're addicted to our endless summer." After all, if one is not allowed to stay up late, what is the point of being an adult?

—NATALIE ANGIER, March 1995

A Mouse Helps Explain What Makes Us Tick

"OH DEAR, oh dear, I shall be too late," the White Rabbit exclaimed as he examined his pocket watch.

The rabbit overheard by Lewis Carroll's Alice in her adventures in Wonderland now has a real-life counterpart: a mouse whose biological clock has been genetically reset to run at a 28-hour cycle instead of 24.

Deranging the daily rhythm of a mouse is not some perverse goal of modern science but rather a means to a larger end, the understanding of the biological clocks that govern the daily lives of everything from microbes to man.

The biological clock in humans governs rhythms like the sleep-wake cycle, the daily ebb and flow of many hormones, and the variations in mental alertness. It resets itself daily according to the amount of light perceived. Without a well-adjusted clock, the human condition can get very jagged, as is evident when changing a work shift or flying through too many time zones for the clock to keep pace.

The biological clock has long defied analysis, in part because the basic rhythmicity is generated at the deepest levels of the cell. In mammals, the master clock is in the suprachiasmatic nucleus of the brain, just above where the optic nerves from each eye cross over. The nucleus, about the size of a pinhead, consists of some 10,000 nerve cells, each containing a tiny clock at the level of its genes and proteins.

Biologists have identified a few individual components of the clocks used by fruit flies and by a microbe called *Neurospora,* which makes bread mold. But possession of the genes for these components did not, as is often the case, help fish out the counterpart genes in mice or humans.

In frustration, a bold new approach was conceived seven years ago by Dr. Joseph S. Takahashi of Northwestern University in Evanston, Illinois. In fact it was a bold old approach, that of trying to find timekeeping genes in mice with the same mass screening method that was used to find the gene in fruit flies in 1971. But mice are far harder to work with than fruit flies.

The method is to serve meals laced with a heavily mutagenic chemical that randomly changes DNA all along the genome. By screening enough fruit flies or mice for defects in their biological clock, the researcher can hope to find one with an aberrant clock, and then rummage around in its genome to see which gene was disrupted.

A large site was built to watch—for a month—hundreds of mice running on wheels. Wheel-running affords an accurate signal of the animals' internal clock since laboratory mice like to take exercise regularly and will start punctually to the minute at the same time each night.

Dr. Takahashi and his team expected to have to screen several thousand mice before hitting on one with a broken clock. But fortune smiled upon their mouse-deranging endeavors, and it was mouse No. 25 that turned up incurably tardy. No other seriously clock-deranged mouse has yet shown up.

"We were incredibly lucky," Dr. Takahashi said.

The mutant mouse had a basic daily rhythm about an hour longer than normal mice, caused by a single defective gene, which Dr. Takahashi named the Clock gene. Descendants bred to have two defective copies of the Clock gene, one from each parent, had a daily rhythm four hours longer than normal.

Dr. Takahashi and his team discovered the clock-deranged mouse three years ago and then set about trying to identify the damaged gene, a snippet of genetic material hiding somewhere within the three billion units of DNA that make up the mouse's genome. The project took a 10-member team three years.

In the journal *Cell*, the researchers report having laid hands on the gene and describe the molecular features of the protein it specifies. Dr. David P. King led efforts to pinpoint the location of the gene. Dr. Marina P. Antoch directed a clever method of proving the gene's function: She fixed the clock in mutant mice by substituting a correct piece of DNA.

Of great interest to specialists is the finding that one stretch of the protein, known as a domain, resembles a domain seen in the clock proteins of the fruit fly and the bread mold, even though the overall protein structures have little similarity. This bolsters the expectation, hitherto unsupported, that since circadian rhythms are an ancient behavior in all living creatures they should have a common basic mechanism.

The common domain is one that enables the protein to activate certain genes strung along the cell's DNA, though the target genes have not yet been identified. Just how this constitutes the mechanism of a clock remains to be determined. From what is known of the fruit fly and bread mold clocks, ingenious schemes have been suggested in which a Clock gene makes a protein, which links up with another protein and then turns off the gene in a daily feedback loop.

Dr. Steven M. Reppert, who studies circadian rhythms at Massachusetts General Hospital, said the Takahashi team's work was "really a tour de force," adding, "This is the first molecular entry into the mammalian clock."

Dr. Charles A. Czeisler, a neuroendocrinologist at Brigham and Women's Hospital in Boston, described the finding as "a landmark discovery which holds great promise for understanding the underpinnings" of the human biological clock.

Some people are night owls, others are alert in the mornings, a difference that may have a genetic basis.

"Given that a Clock gene has now been identified in mice, one can begin to tease apart whether these differences in human behavior may have a genetic basis," Dr. Czeisler said.

Dr. Takahashi's team found that the normal version of the Clock gene in mice is made up of 24 sections, known as exons. In the mutant mouse they created, a single unit of DNA was changed, causing one of the exons to be lost from the processed version of the gene. Loss of the exon resulted in a protein with one part missing, which slowed the clock.

Progress in Clock gene research has been gathering pace after a slow start. The two Clock genes so far known in the fruit fly are called Period, found in 1971, and Timeless, found in 1994. The bread mold gene Frequency was discovered in 1978. Two more bread mold genes, dubbed White collar-1 and White collar-2, were reported earlier this month by a team led by Jay C. Dunlap and Jennifer J. Loros at the Dartmouth Medical School in Hanover, New Hampshire.

—NICHOLAS WADE, May 1997

Scientists Report the Discovery of a Brain "Switch" That Brings on Sleep

SCIENTISTS REPORT that they have found a master switching mechanism for sleep.

When the switch—a tiny clump of cells deep in the brain—is turned on, all brain cells involved in arousal and awareness are shut down. Conversely, when the switch is turned off, the brain wakes up.

Everyone has experienced the switch at work, said Dr. Clifford Saper, chief of neurology and neuroscience at Beth Israel Hospital in Boston, who led the research. You are drowsy. The room is hot, the lecture is boring, and you cannot keep your eyes open. The switch turns on, Dr. Saper said, and you are out.

It is not this mechanism, however, that makes people drowsy and starts them on the somnolent slide. That, Dr. Saper said, is another mechanism, yet to be discovered, a kind of dimmer switch. The newly discovered clump of cells turns out the lights entirely.

The new finding, described in *Science* magazine, was demonstrated in rats, but the researchers said that it almost certainly applied to humans, too. The brain circuits controlling sleep are highly similar in all mammals.

The work looks "very interesting," said Dr. Peter Reiner, an associate professor of neuroscience at the University of British Columbia in Vancouver. "This seems to be a master switch. If we can control it, it opens the door to novel ways of modifying sleep and waking states. And it opens a whole new area of the brain for detailed study."

In practical terms, the finding might lead to new, more efficient medications to treat insomnia and somnolence, Dr. Reiner said.

But it remains to be seen how the switch fits into the whole pattern of sleep, which is a complicated state, said Dr. Adrian Morrison, a sleep researcher at the University of Pennsylvania. "What they found," Dr. Morrison said, "is an important node in a larger system."

Researchers have known that there must be brain cells, or neurons, that turn on only at the onset of sleep. But these cells are not marked with signposts in the brain, Dr. Saper said, and so no one knew where they were or how they were connected to the billions of cells that function when the brain is awake.

Moreover, although there are many small clumps of specialized cells in the brain stem and the forebrain that help orchestrate the daily cycles of wakefulness and sleep—circadian rhythms—it has been a mystery how these cell groups connect to one another, Dr. Saper said, and how a master switch might work.

Clues have been around for a long time, though. During World War II, Dr. Wally Nauta, a Dutch researcher working in Amsterdam, captured street rats and carried out surgery on the hypothalamus, a tiny brain structure that controls many hormones and regulatory functions in the body.

When Dr. Nauta cut the front part of the hypothalamus, the rats could not go to sleep, Dr. Saper said; they were total insomniacs. But when he cut the back part, the rats became comatose. This pointed to the idea that the hypothalamus contained a master switch for sleep, Dr. Saper said, but the exact location and chemical nature of the switch remained unknown.

To discover these details, Dr. Saper and his colleagues conducted a series of experiments on rats, using chemical tracers that identify a protein found in cells when they are active. In the initial experiments, 20 rats in various stages of wakefulness or sleep were killed, and their brain tissue was examined.

The rats that had been awake when they were killed showed the protein, called fos, throughout the brain, Dr. Saper said. But animals that had been sleeping showed fos protein in just three tiny areas of the brain: in a front region of the hypothalamus called the ventrolateral preoptic neurons, or VPN, and in two structures known to regulate circadian rhythms.

To find out whether the VPN was another controller of circadian rhythms or instead a master sleep switch, the researchers carried out a second set of experiments, on 33 rats. The animals were deprived of sleep for 9 to 12 hours and then either killed immediately or allowed to sleep a few hours before being killed.

Those that had not been allowed to sleep showed fos protein throughout most of the brain but not in the VPN The rats that had been allowed to get some sleep showed pronounced amounts of fos in the VPN, but not in the rest of the brain. This showed that fos buildup in the VPN was related not to the need for sleep but to sleeping itself, Dr. Saper said.

Moreover, in the sleep-deprived rats the regions of the brain that govern circadian rhythms did not show any fos activity, because the rats had been killed during their "off cycle"—that is, at a time when their circadian structures were not active, Dr. Saper said.

In humans, the VPN contains 20,000 to 40,000 cells that produce a neurotransmitter, called GABA (γ-aminobutyric acid), that inhibits the firing of other brain cells, Dr. Saper said. Moreover, the VPN sends connections to three groups of specialized cells: one in the back of the hypothalamus and two in the brain stem. These cells literally keep the brain awake, he said, by sending arousal signals via neurotransmitters to every part of the cortex, ultimately affecting billions of cells.

It looks as if the VPN inhibits all the neurotransmitters involved in wakefulness, arousal and consciousness, Dr. Saper said. This, in fact, may be a major function of sleep—to allow these structures time to regenerate.

"In sleep," he said, "the whole system is reset."

—SANDRA BLAKESLEE, January 1996

Mystery of Sleep Yields as Studies Reveal Immune Tie

A COLLEGE STUDENT goes two nights without sleep to cram for exams and on the third day comes down with a cold. A night-shift employee begins working days and gets the flu. A surgery patient who is awakened four times a night in the hospital begins to recover only after going home and getting a good night's sleep.

Are these situations coincidental? Or do they show that sleep loss promotes illness? Despite intense interest in the question, sleep researchers have been hard pressed to show exactly how sleep influences human health and disease.

But now a burst of findings is beginning to shed light on the ultimate purpose of sleep, and in particular on the convoluted interplay between sleep and the immune system. Experiments suggest that the immune system is somehow repaired or bolstered during sleep and that it, in turn, has a role in regulating sleep.

Sleep is divided into periods of so-called REM sleep, characterized by rapid eyeball movements and dreaming, and longer periods of non-REM sleep. Neither kind of sleep is at all well understood, but REM sleep is assumed to serve some restorative function of the brain. The purpose of non-REM sleep is even more mysterious. The new experiments, such as those described for the first time at a recent meeting of the Society for Sleep Research in Minneapolis, suggest intriguing explanations for the purpose of non-REM sleep.

Sleep serves many purposes, said Dr. Harvey Moldofsky, director of the University of Toronto Center for Sleep and Chronobiology. Apparently, animals sleep to regulate body temperature, organize memories and replenish the immune system, he said. But most research has focused on sleep as a brain phenomenon, ignoring the rest of the body. The cells, organs, hormones and immune factors in the periphery may, like the brain, contain molecular clocks that help drive daily sleep and wake cycles, he said.

Dr. James Krueger, a physiologist at the University of Tennessee in Memphis, has investigated the idea that sleep factors—the molecules that promote sleep—build up in the bloodstream during the day and, when they reach a high enough concentration, make people drowsy. Such sleep factors would be only one mechanism for promoting sleep, he said. Others might kick in during an infection, for example, or on a hot summer day or after Thanksgiving dinner. Although the mechanisms of the various sleep factors would be different, he said, they may interact with and compensate for each other.

Dr. Krueger is focusing much of his research on cytokines, messenger chemicals of the immune system that are associated with white blood cells. These substances, which make up the front line against infection, are also found in the brain, although whether circulating cytokines converse with or exert control on brain cytokines is not known.

When cytokines like interleukin-1 (IL-1) and tumor necrosis factor are injected into animal brains, the animals fall asleep, Dr. Krueger said. It may be that these molecules promote sleep in some regions of the brain and not others, so that the whole brain does not sleep at once, he said. In fact, different brain networks may take turns sleeping, which would explain why there are many gradations of sleep, from light dozing to the deepest sleep.

Some of the new research began with a mystery that presented itself 10 years ago. Dr. Allan Rechtschaffen at the University of Chicago Sleep Research Laboratory put two rats into the same environment, but permitted only one to sleep. No striking differences emerged until the end of the second week, when the sleep-deprived rat began to gorge itself on food yet grew skinnier and skinnier without exercising more. After one more week, the sleep-deprived rat lost the ability to regulate its body temperature and died.

Later, in looking for the cause of death of sleep-deprived rats in many such experiments, researchers could find nothing wrong with them. Their organs, blood and urine all seemed normal. The animals resembled cancer patients whose bodies are either weakened by chemotherapy or wasting away from their disease, said Dr. Carol Everson, a senior staff fellow at the National Institute of Mental Health in Bethesda, Maryland, and a former student of Dr. Rechtschaffen's.

"I began thinking, what could be toxic? Maybe the rats were infected," she said. After culturing their blood, Dr. Everson found that the rats had died from bacterial infections of the blood. The bacteria were strains that the animals were in contact with every day, she said, and do not normally cause disease.

Oddly, these infections did not damage tissue, Dr. Everson said, suggesting that the rats' immune systems did not mount an aggressive attack on the bacteria. Further tests are under way to measure the immune response during the sleep-deprivation experiments, she said.

Dr. David Dinges, a psychiatrist at the University of Pennsylvania, is testing the effects of sleep deprivation on healthy men and women, meanwhile. There is a long-held belief, based on very little evidence, that going without sleep will make you sick, Dr. Dinges said. Some studies have shown that medical students taking exams, caretakers of patients with Alzheimer's disease and people in bereavement have reduced lymphocyte counts—the T cells and B cells that combat infection—and decreases in other immune system cells, he said. The idea is that people in crisis who may not be sleeping well have depressed immune systems.

But few studies have looked at how healthy people respond to sleep loss, Dr. Dinges said. In an ambitious experiment that is now being analyzed, Dr. Dinges and his colleagues recruited 24 healthy volunteers who agreed to live in a sleep lab for one week. Dr. Dinges said that he and others expected to find a decline in immune function after sleep loss. "But from the git-go," he said, "we realized we were on to something different."

The T cells and B cells that are called upon to attack specific pathogens showed no change, he said. But monocytes, granulocytes and natural killer cells—immune cells that are called into play when the body responds to an unknown invader—went sky-high. Cytokines like IL-1 also appear to be elevated, he said.

The sleep-deprived subjects seem to be mounting what immunologists call a non-specific host response, Dr. Dinges said. It is a first line of defense against disease-causing agents and means that these people should be better at fighting off colds and flu. Whether this response would endure after additional hours of sleep deprivation is not known, he said. And what it means to the brain is also not known.

Such experiments raise the question of what role the immune system plays in normal sleep. One idea is that immune cells are involved in the buildup of sleep factors.

In 1980, Dr. Krueger helped identify one such factor in the human gut. During the day, he said, immune cells called macrophages digest gut bacteria and release tiny proteins from the bacterial cell walls. These proteins escape into the circulation and may reach the brain. When injected into animal brains, the proteins produce deep prolonged sleep.

In addition, Dr. Krueger said, the macrophages stimulate the release of cytokines. They, too, induce sleep when injected into animal brains.

Such findings may also explain why people get sleepy when they are sick, Dr. Krueger said. Cytokines and other immune cells flourishing in the blood and lymph may exert an influence on the brain, inducing sleep while the infection lasts.

Dr. Krueger, a pioneer in the field of sleep and the immune system, has distilled his observations into a theory of sleep. The brain is composed of myriad groups of neurons that carry out specific functions, he said. But during the day, not every group is called into play. If the neuronal groups are not stimulated, their connections may be lost. During sleep, Dr. Krueger said, the brain releases cytokines that induce a special firing pattern among various neuronal groups, preserving their connections for future use.

Thus parts of the brain sleep while other parts are awake, Dr. Krueger said. The collective output of many such groups leads to what scientists called non-REM sleep, when dreams do not occur.

Such research has major implications for human ailments. Chronic fatigue syndrome could be related to the abnormal arousal of cytokines in the brain. Cancer patients and transplant patients are prone to developing bacterial infections of the blood similar to those seen in the sleep-deprived rats. The practice of taking the temperature and blood samples of hospitalized patients throughout the night could, in some cases, do more harm than good. And AIDS patients could be exhausted during the day because their damaged immune systems interfere with normal sleep.

Recent experiments with AIDS patients and sleep suggest ways in which factors outside the brain might help regulate sleep, said Dr. Dennis Darko, a psychiatrist at the Scripps Clinic and Research Foundation in La Jolla, California.

When healthy people sleep, the intensity of a characteristic sleep wave, called delta, goes up and down throughout the night, Dr. Darko said. Moreover, a cytokine called tumor necrosis factor undergoes a synchronous fluctuation.

This synchrony is off in AIDS patients, Dr. Darko said, and may be a clue as to why patients are so tired during the day.

But the experiment raises a larger question, he said. Why would healthy people have waxing and waning levels of tumor necrosis factor in their blood? This would indicate a periodic activation of the immune system throughout the night, but what might kick it off?

"We asked the gastroenterologists if they could think of anything," Dr. Darko said, and they had an answer. About four hours after the last meal of the day and sleep onset, the small bowel, which must remain sterile, begins chugging with rhythmic contractions. A slow wave starts at the bottom of the stomach and for the next 90 to 120 minutes moves down through the small bowel to the junction between the small and large intestine. Bacteria are pushed away from the small bowel to the colon, which is built to handle them, Dr. Darko said.

All night long, he said, the colon's walls are flooded by churning bacteria in rhythms lasting 90 to 120 minutes—which matches exactly the interval between the sleep stages seen in the brain.

It could be a coincidence, Dr. Darko said. But it could also mean sleep factors are reaching the brain in synchronous waves. REM sleep, when dreams occur, may be necessary for a healthy brain, he said. "But for all we know, non-REM sleep may be necessary for a healthy bowel."

—SANDRA BLAKESLEE, August 1993

7

SEXUAL DIFFERENCES IN THE BRAIN

Men and women have the same basic body plan, but one that is sculpted in different ways by the hormones released in the developing embryo.

There has long been evidence from animals that the brain also is shaped during development into male and female versions. Sexual differences also exist in the human brain, though they seem to be more subtle than the bodily differences between men and women. The field is one that has harbored many false assumptions in the past and it is hard for even the most careful scientists to persuade others that their results are perfectly objective.

The findings point toward the view that male and female brains are specialized, as is the body, to perform somewhat different roles, although the overlap is far greater than the differences.

The field has assumed a new complexity with the claims that sexual preferences may be genetically based and associated with structural differences in the brain. But this is a young field in which few claims have been put beyond dispute.

Man's World, Woman's World? Brain Studies Point to Differences

DR. RONALD MUNSON, a philosopher of science at the University of Missouri, was elated when *Good Housekeeping* magazine considered publishing an excerpt from the latest of the novels he writes on the side. The magazine eventually decided not to publish the piece, but Dr. Munson was much consoled by a letter from an editor telling him that she liked the book, which is written from a woman's point of view, and could hardly believe a man had written it.

It is a popular notion: that men and women are so intrinsically different that they literally live in different worlds, unable to understand each other's perspectives fully. There is a male brain and a female brain, a male way of thinking and a female way. But only now are scientists in a position to address whether the notion is true.

The question of brain differences between the sexes is a sensitive and controversial field of inquiry. It has been smirched by unjustifiable interpretations of data, including claims that women are less intelligent because their brains are smaller than those of men. It has been sullied by overinterpretations of data, like the claims that women are genetically less able to do everyday mathematics because men, on average, are slightly better at mentally rotating three-dimensional objects in space.

But over the years, with a large body of animal studies and studies of humans that include psychological tests, anatomical studies and, increasingly, brain scans, researchers are consistently finding that the brains of the two sexes are subtly but significantly different.

Now, researchers have a new non-invasive method, functional magnetic resonance imaging, for studying the live human brain at work. With it, one group recently detected certain apparent differences in the way men's and women's brains function while they are thinking. While stressing extreme caution in drawing conclusions from the data, scientists say nonetheless that the groundwork was being laid for determining what the differences really mean.

"What it means is that we finally have the tools at hand to begin answering these questions," said Dr. Sally Shaywitz, a behavioral scientist at the Yale University School of Medicine. But she cautioned: "We have to be very, very careful. It behooves us to understand that we've just begun."

The most striking evidence that the brains of men and women function differently came from a recent study by Dr. Shaywitz and her husband, Dr. Bennett A. Shaywitz, a neurologist, who is also at the Yale medical school. The Shaywitzes and their colleagues used functional magnetic resonance imaging to watch brains in action as 19 men and 19 women read nonsense words and determined whether they rhymed.

In a paper, published in *Nature,* the Shaywitzes reported that the subjects did equally well at the task, but the men and women used different areas of their brains. The men used just a small area on the left side of the brain, next to Broca's area, which is near the temple. Broca's area has long been thought to be associated with speech. The women used this area as well as an area on the right side of the brain. This was the first clear evidence that men and women can use their brains differently while they are thinking.

Another recent study, by Dr. Ruben C. Gur, the director of the brain behavior laboratory at the University of Pennsylvania School of Medicine, and his colleagues, used magnetic resonance imaging to look at the metabolic activity of the brains of 37 young men and 24 young women when they were at rest, not consciously thinking of anything.

In the study, published in the journal *Science,* the investigators found that for the most part, the brains of men and women at rest were indistinguishable from each other. But there was one difference, found in a brain structure called the limbic system that regulates emotions. Men, on average, had higher brain activity in the more ancient and primitive regions of the limbic system, the parts that are more involved with action. Women, on average, had more activity in the newer and more complex parts of the limbic system, which are involved in symbolic actions.

Dr. Gur explained the distinction: "If a dog is angry and jumps and bites, that's an action. If he is angry and bares his fangs and growls, that's more symbolic."

Dr. Sandra Witelson, a neuroscientist at McMaster University in Hamilton, Ontario, has focused on brain anatomy, studying people with terminal cancers that do not involve the brain. The patients have agreed to participate in neurological and psychological tests and then to allow Dr. Witelson and her colleagues to examine their brains after they die, to look for relationships between brain structures and functions. So far she has studied 90 brains.

Several years ago, Dr. Witelson reported that women have a larger corpus callosum, the tangle of fibers that run down the center of the brain and enable the two hemispheres to communicate. In addition, she said, she found that a region in the right side of the brain that corresponds to the region women used in the reading study by the Shaywitzes was larger in women than in men.

Most recently, Dr. Witelson discovered, by painstakingly counting brain cells, that although men have larger brains than women, women have about 11 percent more neurons. These extra nerve cells are densely packed in two of the six layers of the cerebral cortex, the outer shell of the brain, in areas at the level of the temple, behind the eye. These are regions used for understanding language and for recognizing melodies and the tones in speech. Although the sample was small, five men and four women, "the results are very, very clear," Dr. Witelson said.

Going along with the studies of brain anatomy and activity are a large body of psychological studies showing that men and women have different mental abilities. Psychologists have consistently shown that men, on average, are slightly better than women at spatial tasks, like visualizing figures rotated in three dimensions, and women, on average, are slightly better at verbal tasks.

Dr. Gur and his colleagues recently looked at how well men and women can distinguish emotions on someone else's face. Both men and women were equally adept at noticing when someone else was happy, Dr. Gur found. And women had no trouble telling if a man or a woman was sad. But men were different. They were as sensitive as women in deciding if a man's face was sad—giving correct responses 90 percent of the time. But they were correct about 70 percent of the time in deciding if women were sad; the women were correct 90 percent of the time.

"A woman's face had to be really sad for men to see it," Dr. Gur said. "The subtle expressions went right by them."

Studies in laboratory animals also find differences between male and female brains. In rats, for example, male brains are three to seven times larger than female brains in a specific area, the preoptic nucleus, and this difference is controlled by sex hormones that bathe rats when they are fetuses.

"The potential existence of structural sex differences in human brains is almost predicted from the work in other animals," said Dr. Roger Gorski, a professor of anatomy and cell biology at the University of California in Los Angeles. "I think it's a really fundamental concept and I'm sure, without proof, that it applies to our brains."

But the question is, if there are these differences, what do they mean?

Dr. Gorski and others are wary about drawing conclusions. "What happens is that people overinterpret these things," Dr. Gorski said. "The brain is very complicated, and even in animals that we've studied for many years, we don't really know the function of many brain areas."

This is exemplified, Dr. Gorski said, in his own work on differences in rat brains. Fifteen years ago, he and his colleagues discovered that males have a comparatively huge preoptic nucleus and that the area in females is tiny. But Dr. Gorski added: "We've been studying this nucleus for 15 years, and we still don't know what it does. The most likely explanation is that it has to do with sexual behavior, but it is very, very difficult to study. These regions are very small and they are interconnected with other things." Moreover, he said, "nothing like it has been shown in humans."

And, with the exception of the work by the Shaywitzes, all other findings of differences in the brains or mental abilities of men and women have also found that there is an amazing degree of overlap. "There is so much overlap that if you take any individual man and woman, they might show differences in the opposite direction" from the statistical findings, Dr. Gorski said.

Dr. Munson, the philosopher of science, said that with the findings so far, "we still can't tell whether the experiences are different" when men and women think. "All we can tell is that the brain processes are different," he said, adding that "there is no Archimedean point on which you can stand, outside of experience, and say the two are the same. It reminds me of the people who show what the world looks like through a multiplicity of lenses and say, 'This is what the fly sees.'" But, Dr. Munson added, "We don't know what the fly sees." All we know, he explained, is what we see looking through those lenses.

Some researchers, however, say that the science is at least showing the way to answering the ancient mind-body problem, as applied to the cognitive worlds of men and women.

Dr. Norman Krasnegor, who directs the human learning and behavior branch at the National Institute of Child Health and Human Development, said the difference that science made was that when philosophers talked about mind, they "always were saying, 'We've got this black box.'" But now, he said, "we don't have a black box; now we are beginning to get to its operations."

Dr. Gur said science was the best hope for discovering whether men and women inhabited different worlds. It is not possible to answer that question simply by asking people to describe what they perceive, Dr. Gur said, because "when you talk and ask questions, you are talking to the very small portion of the brain that is capable of talking." If investigators ask people to tell them what they are thinking, "that may or may not be closely related to what was taking place" in the brain, Dr. Gur said.

On the other hand, he said, scientists have discovered that what primates perceive depends on how their brains function. Some neurons fire only in response to lines that are oriented at particular angles, while others seem to recognize faces. The world may well be what the philosopher Descartes said it was, an embodiment of the workings of the human mind, Dr. Gur said. "Descartes said that we are creating our world," he said. "But there is a world out there that we can't know."

Dr. Gur said that at this point he would hesitate to baldly proclaim that men and women inhabit different worlds. "I'd say that science might be leading us in that direction," he said, but before he commits himself he would like to see more definite differences in the way men's and women's brains function and to know more about what the differences mean.

Dr. Witelson cautioned that "at this point, it is a very big leap to go from any of the structural or organizational differences that were demonstrated to the cognitive differences that were demonstrated." She explained that "all you have is two sets of differences, and whether one is the basis of the other has not been shown." But she added, "One can speculate."

Dr. Witelson emphasized that in speculating she was "making a very big leap," but she noted that "we all live in our different worlds and our worlds depend on our brains.

"And," she said, "if these sex differences in the brain, with 'if' in big capital letters, do have cognitive consequences, and it would be hard to believe there would be none, then it is possible that there is a genuine difference in the kinds of things that men and women perceive and how these things are integrated. To that extent it may be possible that in some respects there is less of an easy cognitive or emotional communication between the sexes as a group because our brains may be wired differently."

The Shaywitzes said they were reluctant even to speculate from the data at hand. But, they said, they think that the deep philosophical questions about the perceptual worlds of men and women can eventually be resolved by science.

"It is a truism that men and women are different," Dr. Bennett Shaywitz said. "What I think we can do now is to take what is essentially folklore and place it in the context of science. There is a real scientific method available to answer some of these questions."

Dr. Sally Shaywitz added: "I think we've taken a qualitative leap forward in our ability to ask questions." But, she said, "the field is simply too young to have provided more than a very intriguing appetizer."

—GINA KOLATA, February 1995

Study Links Brain to Transsexuality

A STRUCTURE deep within the brain, where the tangled roots of sexuality are thought to lie, differs substantially between ordinary men and transsexuals who have surgically transformed themselves from men into women, scientists have reported.

Researchers in the Netherlands have discovered that a region of the hypothalamus, located at the floor of the brain, is about 50 percent larger in men than in women, and almost 60 percent larger in men than in male-to-female transsexuals. If smallness of this brain structure is at all correlated with the feeling of being a woman, the results raise tantalizing possibilities that transsexuals may in a sense be more female than females.

The discovery is the first detection of a difference in transsexual brains and could at least partly explain why such individuals describe themselves as "women trapped in men's bodies."

The finding may also cast light on the larger issue of sexual identity, of what makes a person feel comfortable—or tormented—in the skin of a man or a woman.

Significantly, the region of the hypothalamus does not differ in size between gay and straight men, and so it cannot be said to play a role in male sexual orientation. Other recent studies have focused on identifying minor brain discrepancies between homosexual and heterosexual men, in general reporting that gay brains appeared comparatively feminine. Such findings, which remain deeply contested, have troubled many people for the simple reason that gay men overwhelmingly think of themselves as men, not as abnormal women. But genetic men who undergo sex reassignment often claim that they felt like girls from early childhood on.

Dr. Dick F. Swaab of the Netherlands Institute for Brain Research in Amsterdam, who with his colleagues is reporting the work in the journal

Nature, emphasized that this section of the hypothalamus is by no means the entire source of sexual identity.

"I'm convinced this is only one structure of many that are involved in such a complex behavior," he said. "This is just the tip of the iceberg."

In addition, the study remains to be replicated by other researchers, which will not be easy. It was performed by dissecting the autopsied brains of transsexuals, homosexual men, heterosexual men and heterosexual women. Because transsexuality is rare, it took the scientists 11 years to collect six transsexual brains.

But while the number of transsexual brains examined is small, Dr. Swaab said the results had scientific power because the discrepancies in size of the hypothalamic structure were quite large.

Others in the field concur. "These are astonishing data," said Dr. Geert de Vries, a neurobiologist at the University of Massachusetts at Amherst.

The specific region under study is called the central subdivision of the bed nucleus of the stria terminalis. Research in rats and other laboratory animals indicates that this part of the hypothalamus helps coordinate sexual behavior and the release of essential reproductive hormones.

But Dr. S. Marc Breedlove of the University of California at Berkeley, who wrote an editorial that accompanies the new report, said that the function of the bed nucleus in human behavior, sexual or otherwise, remained "a complete black box."

In heterosexual and homosexual men, the bed nucleus measures about 2.6 cubic millimeters, about the size of the colorful, spherical head of a pushpin. In women, it averages 1.73 cubic millimeters, and in transsexuals the average figure is 1.3.

Some scientists cautioned that the estrogen treatment the transsexuals took as part of their sex-change therapy might have affected the size of their hypothalamus; but the Dutch researchers tried to rule out that factor by including brains of transsexuals who had stopped estrogen years earlier, as well the brains of men and women with varying hormonal conditions. In no case did the size of the bed nucleus appear to be influenced by adult hormone levels.

The researchers are now trying to collect autopsied brains from women who were surgically changed to men—a considerably less common type of transsexuality than male-to-female—to see if their hypothalamic regions are of masculine dimensions.

Some scientists remain guarded about the meaning of the latest work, as do people in the increasingly vocal "transgendered" community, which includes transsexuals and those born with conditions like hermaphroditism, in which both male and female sexual organs are present. "I think that these studies have a very poor history of reproducibility, and that they're undertaken with a particular social agenda," said Cheryl Chase, editor of *Hermaphrodites With Attitude,* the newsletter of the Intersex Society of North America.

But Dr. Joy Diane Shaffer, director of the Seahorse Medical Clinic in San Jose, California, and herself a transsexual, said the results jibed with those that she and her colleagues were gathering, using a very different approach to brain analysis. They are using magnetic resonance imaging technology to scan the brains of hundreds of living people, including heterosexuals of both sexes and transsexuals of both directions. Their method would not pick up differences in the tiny bed nucleus, but it may observe differences in other, larger structures, like the corpus callosum, which connects the two hemispheres of the brain.

"The social pressure to conform to the role of birth assignment is so strong, and the penalties for changing gender as an adult are so great, that you'd expect people who make the role switch to be substantially transgendered" in the brain, said Dr. Shaffer. She estimates that there are perhaps 10,000 male-to-female transsexuals in the United States. Many more might become so if they could afford it, she added, and in fact transsexuals often do seem to be among the socioeconomically advantaged. There are a number of famous transsexuals, including the tennis player Rene Richards, the model Tula and the writer Jan Morris. Transsexual sexuality is diverse, with some preferring male partners, and others choosing women as lovers.

Dr. Swaab proposes that the sexual variances in the size of the bed nucleus arise during fetal development, and thus are essentially built in. But Dr. Roger Gorski, a neurobiologist at the University of California at Los Angeles, said the possibility could not be ruled out that the changes in the hypothalamus occurred after birth, perhaps as a result of one's behavior while growing up, or even during early adolescence when a surge in sex hormones flooded the brain.

—NATALIE ANGIER, November 1995

Report Suggests Homosexuality Is Linked to Genes

USHERING the politically explosive study of the origins of sexual orientation into a new and perhaps more scientifically rigorous phase, researchers report that they have linked male homosexuality to a small region of one human chromosome.

The results have yet to be confirmed by other laboratories, and the chromosomal region implicated, if it holds up under further scrutiny, is almost surely just a single chapter in the intricate story of sexual orientation and behavior. Nevertheless, scientists said the work suggests that one or several genes located on the bottom half of the sausage-shaped X chromosome may play a role in predisposing some men toward homosexuality. (The researchers have begun a similar study looking at the chromosomes of lesbians.)

The findings, which appear in the journal *Science,* indicate that sexual orientation often is at least partly inborn, rather than being solely a matter of choice. But researchers warn against overinterpreting the work, or in taking it to mean anything as simplistic as that the "gay gene" had been found.

The researchers emphasized that they do not yet have a gene isolated, but merely know the rough location of where the gene or genes may sit amid the vast welter of human DNA. Until they have the gene proper, scientists said they have no way of knowing how it contributes to sexual orientation, how many people carry it, or how often carriers are likely to become gay as a result of bearing the gene.

And even when they do have this gene on the X chromosome pinpointed, scientists said they will continue to search for other genes on other chromosomes that may be involved in sexual orientation.

"Sexual orientation is too complex to be determined by a single gene," said Dr. Dean H. Hamer of the National Cancer Institute in Bethesda, Maryland, the lead author on the new report. "The main value of this work is that it opens a window into understanding how genes, the brain and the environment interact to mold human behavior."

In the new work, the scientists studied the genetic material from 40 pairs of gay brothers and found that in 33 of the pairs, the brothers had identical pieces of the end tip of the X chromosome. Under ordinary circumstances ruled by chance alone, only half of the pairs should have shared that chromosomal neighborhood in common, a region designated Xq28. The odds of Dr. Hamer's results turning up randomly are less than half a percent, indicating that the chromosomal tip likely harbors a genetic sequence linked to the onset of the brothers' homosexuality.

In men, the X chromosome pairs with the Y chromosome to form the so-called sex chromosomes, the final set of the 23 pairs of chromosomes found in all cells of the human body. A man's X chromosome is always inherited from the mother, who bestows on her son a reshuffled version of one of her two copies of the X chromosome. The latest results indicate that the newly reported genetic factor is passed through the maternal line, a curious twist given that in the past psychiatry has held women at least partly responsible for fostering their sons' homosexuality.

The gene could work by directly influencing sexual proclivity, perhaps by shaping parts of the brain that orchestrate sexual behavior. Or it might affect temperament in a way that predisposes a boy toward homosexuality. At the moment, researchers said, all scenarios are mere speculation.

Despite the cautionary notes, the latest study is likely to add fuel to the debate over gay rights in the military and civilian realms.

If homosexuality is shown to be largely inborn, a number of legal experts say, then policies that in any way discriminate against homosexuals are likely to be shot down in the courts.

"We think this study is very important," said Gregory J. King, a spokesman for the Human Rights Campaign Fund in Washington, the largest national lesbian and gay lobbying group. "Fundamentally it increases our understanding of the origins of sexual orientation, and at the same time we believe it will help increase public support for lesbian and gay rights."

Not all gay rights leaders have a sanguine view of the work. Some denounce it as yet another attempt to draw a reductionist and implacable line between homosexuality and heterosexuality, while others see in it the dangers of attempts to "fix" homosexuality, perhaps through gene therapy.

"I don't think it's an interesting study," said Darrell Yates Rist, co-founder of the Gay and Lesbian Alliance Against Defamation. "Intellectually, what do we gain by finding out there's a homosexual gene? Nothing, except an attempt to identify those people who have it and then open them up to all sorts of experimentation to change them."

The study appears in the same journal that two years ago unleashed a furious debate when it published a report asserting to have found an anatomical difference between the brains of gay and straight men. Other recent reports have also weighed in on the possible biological basis of homosexuality in both men and women, and all have been subjected to volleys of scientific and political attack.

Other attempts to make a genetic link to behavior, like alcoholism, manic depression and schizophrenia, have thus far all been disappointing. By contrast, today's study is considered to be impressive science even by many who denounced the previous studies.

"It's a good piece of work," said Dr. Anne Fausto-Sterling of Brown University, a geneticist who has been one of the most outspoken critics of the genetic studies of human behavior. "Hamer is appropriately cautious about the meanings you can glean from it, and he admits that there may be more than one path to the endpoint of homosexuality."

Dr. Eric Lander of the Whitehead Institute for Biomedical Research in Cambridge, Massachusetts, said, "From a geneticist's point of view, if you strip away the non-scientific considerations, it's an interesting finding that merits being followed up in a larger sample."

Dr. Simon LeVay, chairman of the Institute of Gay and Lesbian Studies in West Hollywood, California, who did the 1991 study comparing gay and heterosexual brains, was considerably more ecstatic in his appraisal of the latest report.

"For so long people have been thinking there is some genetic element for sexual orientation and this is by far and away the most direct evidence there has been," Dr. LeVay said. "It's the most important scientific finding ever made in sexual orientation."

In the latest experiments, the researchers began by taking the family histories of 114 men who identified themselves as homosexual. Much to their surprise, the researchers discovered a higher-than-expected number of gay men among the men's maternal uncles and male cousins who were the sons of their mothers' sisters. The ratio was far higher than for men in the general population, suggesting a gene or genes that is passed through the maternal line and thus through the X chromosome.

The scientists then focused on gay brothers, on the assumption that if two boys in a family are homosexual, they were more likely to be so for genetic reasons than were those homosexual men without gay brothers. Using a well-known genetic technique called linkage mapping, they scrutinized the X chromosome in the brothers and other relatives by the application of DNA markers, tiny bits of genetic material that can distinguish between chromosomes from different people. The researchers found that more than three quarters of the brothers had inherited identical DNA markers on the Xq28 region of the chromosome.

"I was surprised at how easy this was to detect," Dr. Hamer said. "Part of that ease was because we were working with the X chromosome," the most extensively studied chromosome in the human genetic blueprint. So far the study has been limited to men who said they were gay, eliminating the ambiguity that would come from considering the genes of men who called themselves heterosexual.

Nonetheless, the region is about four million bases, or DNA building blocks long, and hence holds hundreds of genes, meaning the scientists have much work ahead of them to sort out which gene or genes is relevant. The researchers are also trying to perform a similar linkage study on lesbian sisters, but so far they have not managed to find a chromosomal region that is consistently passed along in families.

—NATALIE ANGIER, July 1993

8

THE NATURE OF DREAMS AND CONSCIOUSNESS

The brain, to use a phrase of neuroscientist Rodolfo Llinas, is an image-making machine. In its default mode it generates the random images that we call dreams. In waking mode, the images are modulated by incoming sensory data into precise representations of the external world.

The relationship of these images to consciousness is a matter of interest to philosophers as well as neurophysiologists. Some believe that once we understand how the images are generated—a formidable problem but one that seems to be within reach—we will also understand everything we need to know about consciousness.

Others argue that consciousness is more complicated than mere perception; we not only see the world, we feel it, and these qualitative sensations cannot be explained by the mechanics of the brain's operation.

Steady progress in neurophysiology has provided much of the basis for the new wave of speculation about consciousness. But as the following chapters make evident, there are still far more questions than answers.

Nerve Cell Rhythm May Be Key to Consciousness

DON'T THINK of a hippopotamus.

Now that you did, think about where this image formed in your mind. How did it come into being? Do mental events, like this hippo, arise from the firing of large sets of brain cells or from something less tangible? What is the stuff of self-awareness and the nature of consciousness?

This question—how does the brain break down and recombine information from the outside world?—is called the binding problem. Once shunned as too inaccessible, it is a hot topic at the annual meeting of the Society of Neuroscience being held in California.

Decades of research on visual perception have led scientists to the finding that the brain first breaks down the elements of an image into components that are separately processed. The central problem is how these separate components are reconstituted for the conscious mind.

Consider the example of person looking at an approaching city bus. The image of the bus enters the eye and is immediately processed by nerve cells in the brain that respond to different features of the image. Some cells are activated by color. Others detect the bus's edges, contours, depth, texture and contrasting shadows. Other nerve cells respond to specific kinds of motion—up, down, forward, back—and the speed of motion.

A similar division of labor occurs in the brain's other sensory regions. For hearing, the sound of the bus is broken down in the brain's auditory cortex by different groups of cells that respond to pitch, volume, frequency, direction and other attributes of sound. The smell of the bus is detected by sets of cells in the brain's olfactory bulb. Were the observer to put a hand on the bus, the brain would again analyze touch, joint movement and skin sensation through responses by different groups of cells.

The scene is thus parsed and analzyed, with the disaggregated elements of the bus's image moving through higher and higher regions of the cortex, calling on memory and experience for context, until it somehow comes together in the mind's eye as a complete concept—a moving, smelly, brightly colored object with a familiar role.

But where does the concept all come together?

People used to suppose there was a little man or his equivalent who made sense of the inputs. Some sought a grandmother cell—a single neuron that would fire when seeing the face of one's grandmother. Others searched for the counterpart of a Cinemascope screen in the brain, a place where the separate streams of sensory information are projected to form a complete image.

But who would be watching the screen? Most neuroscientists now agree that there is no little man, no grandmother cell, no theater screen. Hence the binding problem. Where or how are the separate pieces of information bound together?

Several proposed solutions are being debated at the meeting. They invoke a range of concepts like synchrony of oscillation, re-entry networks, chaos, convergence zones and the notion that there really are anatomical sites that give unity to perception.

A leading contender, though highly controversial, is the oscillation theory.

The idea is that separate populations of cells—those responding to aspects of the color, shape, texture, motion, smell and sound of the bus along with those holding memories of buses experienced in the past—all send out nervous impulses at the same firing rate or frequency for a fraction of a second. As they all fire or "oscillate" at this critical frequency, the perception of the bus is created in a network.

So binding, according to this theory, is a matter not of where but of when. Consciousness is but a stream of oscillating networks, forming and falling away every 50 to 100 milliseconds.

The brain's use of frequency as the means to integrate separate parts of a perception has obvious advantages, said Dr. Charles Gray, a neuroscientist at the Salk Institute and early proponent of the oscillation theory. If visual objects were mapped in space, he said, the brain would need special sites for every possible image, whether real or imagined—cows, purple cows, purple cows with blue spots, purple cows with blue spots flying stealth bombers and so on ad infinitum.

"The number of objects we see in a lifetime exceeds the number of neurons that could code information that way," Dr. Gray said. "With a temporal code you can have cells that respond to certain features interacting with cells that combine with other features."

In 1986, Dr. Gray and Dr. Wolf Singer of the Max Planck Institute in Germany found the first evidence of a temporal binding code in animals. Inserting many recording electrodes into the brains of an anesthetized cat, they discovered that sets of neurons at sites widely separated in the brain would fire in synchrony with each other in response to a particular stimulus. The synchronized firing was 40 times a second.

Dr. Singer, Dr. Gray and their co-workers proposed at the time that the 40-cycle-a-second frequency might be characteristic of binding. After repeating the experiment in monkeys, they have modified their theory.

"We still think oscillations are important," Dr. Gray said in an interview this week. "But it's turning out that synchrony may be more important." Synchrony, he said, is when nerve cells fire simultaneously but the interval between joint firing can vary. Such cells are presumably connected by synapses, directly or indirectly.

At the University of Iowa two neurologists, Dr. Antonio Damasio and Dr. Hanna Damasio, who are married, are developing a theory of consciousness that is compatible with the synchrony and oscillation theory. In their view, binding takes place in a hierarchy of anatomical sites call convergence zones. Streams of information are combined in lower-level zones and passed to higher and higher zones depending on the complexity of the task. Consciousness occurs when the higher convergence zones fire signals back to the lower levels, so that the whole architecture "lights up" in synchrony, Dr. Antonio Damasio said.

Dr. Christof Koch of the California Institute of Technology uses a similar analogy of light and levels of processing. "The brain is like a Christmas tree with 10 billion electric candles," he said. "When we pay attention, 10 thousand synchronize and flicker all at once for 100 milliseconds. Then they desynchronize and the next 10 thousand come on."

For many tasks, even fairly complicated ones, the brain binds information unconsciously, Dr. Koch said. Most people have had the experience of being lost in thought while driving home from work and wonder, on arrival, how they got there in one piece, he said. Binding at lower levels, at subconscious levels, probably accounts for automatic, unconscious activities, like walking.

But vivid consciousness may require a different level of binding, involving attention and short-term memory, Dr. Koch said. The synchronous or oscillatory firing of neurons may release chemicals that produce short-term memories, he said, "allowing us to hold five or six things in our heads at once and to have the sense of being conscious."

Dr. Gerald Edelman, director of the Neuroscience Institute at the Scripps Research Foundation in La Jolla, California, is using a parallel supercomputer to test a model of the brain based on similar concepts. Although he uses different terms like correlation and re-entry, his findings support the general notion of synchrony. When given a stimulus, key "neurons" in the model fire in phase whereas millions of background "neurons" have a different phase.

"The model does some exquisite things," Dr. Edelman said, including mimicry of optical illusions that trick the human eye and brain.

This week scientists are presenting evidence that both supports and refutes these ideas.

At the University of Washington in Seattle, Dr. Eberhard Fetz found that when a monkey sits quietly, it has no brain oscillations. But when it has a challenging motor task, like feeling for raisins in the hands of a researcher hidden behind a screen, oscillations arise in two brain areas—one that serves the hand and one where movements are carried out. Oscillations seem to arise, he said, when the animal has to carry out movements requiring attention and fine control.

At Brown University, Dr. John Donoghue found that when a monkey is preparing to make a purposeful move, its brain shows oscillations. But when it carries out the movement, the oscillations fade.

At New York University, Dr. Rodolfo Llinas has measured 40-cycle-per-second signals in humans using a magnetic brain imaging device. "It is a very powerful wave that sweeps from the front to back of the brain every 12.5 milliseconds," he said. "We propose this is a coherent wave which puts together all that happens in the brain at that time. It is macroscopic binding."

But Dr. William Newsome of Stanford University said that he had carried out hundreds of experiments measuring brain activity in awake monkeys and found no evidence of oscillations. While oscillations may exist, he said: "I am very skeptical that they are important for perceptual binding in ways portrayed. This story is being oversold in the face of paltry and conflicting data."

At New York University, Dr. Anthony Movshon found that visually induced nerve cell firings at a rate of 40 per second do not, as might be expected, affect human perception. "I'm not convinced we have to stick everything together with some kind of neural glue," he said. "These different streams may communicate with each other all along in ways we don't understand."

Other neuroscientists are developing alternative theories of binding and consciousness.

At the University of California at Berkeley, Dr. Walter Freeman has a theory based on chaos and networks. "Just like the physics of a water molecule does not tell you about hurricanes," he said, "the property of a single neuron does not shed light on consciousness." Consciousness arises, he said, from cooperative states among networks of neurons.

Dr. David Van Essen and Dr. Charles Anderson of Washington University in St. Louis have an altogether different view. "We think when you need to associate two features or more relating to the world, you need to bring those two types of information to the same locus," Dr. Van Essen said. "It is place, not time, that binds." They propose a dynamic routing scheme in which an area called the pulvinar nucleus acts as a consciousness command post.

No one knows whose ideas will turn out to be correct, said Dr. A. B. Bonds, a professor of electrical engineering at Vanderbilt University who is modeling neuron behavior in computer networks. "We may hold the secret to how the brain works in our hands today," he said. "But we just don't know it's there."

—SANDRA BLAKESLEE, October 1992

Clues to the Irrational Nature of Dreams

FIRST COMES DROWSINESS and a sense that it's time to rest. Eyelids grow heavy. Stray thoughts flicker through the mind as sleep begins, often with a sudden twitch. And then the human brain falls into a state of profound madness filled with hallucinations, delusions and confabulations.

Dreams unfold. We walk, run, fly and float through strange landscapes. Characters appear and turn into different people. Objects are transformed. A rope becomes a snake. Uncle Harry turns into a Tibetan monk and it all makes sense in some screwy, dream-like way.

The bizarre nature of dreams is beginning to make sense to scientists who study the biological and physiological changes that occur in the brain during sleep, wakefulness and the many related states, like dreamless sleep, daydreaming and, some say, the writing of poetry and other creative acts.

For example, researchers have found that during sleep the brain is bombarded by wild, erratic pulses from the brain stem and flooded with nervous system chemicals that induce the insanity of dreams. Areas that control sleep are near areas that control body movements, which explains why eyelids grow heavy with drowsiness and why dreams are full of fictive movements.

Finally, the research is shedding light on the biological basis of some forms of schizophrenia and may also help explain why deep meditation or isolation tanks can induce hallucinations and offer insight into the nature of consciousness itself.

"We study sleeping to understand waking," said Dr. Allan Hobson, a neuroscientist at Harvard University who has helped develop some of the leading theories on the biology of dreams. "We study dreaming to understand madness."

Dr. Hobson and his colleagues presented their latest findings at the annual meeting of the Society for Sleep Research in Boston last month.

Dream researchers disagree about many of the details, "but the Hobson model is the best thing we have going for us now," said Dr. Stanley Krippner, president of the Association for the Study of Dreams and a veteran dream psychologist. "It can tell us a lot about how memory is constructed and reconstructed and how people use personal myths" to define reality.

Machines like electroencephalograms that measure electrical brain activity and others that measure magnetic brain activity cannot distinguish the awake brain from the dreaming brain, Dr. Hobson said in a recent interview.

The awake brain receives copious amounts of information from the outside world, mainly in the form of light and sound frequencies, chemical signals and physical touch, Dr. Hobson said. It processes these signals in vast, oscillating networks of brain cells to form representations of the external world and combines these maps with memories, movements, emotions and forethought in a way that gives rise to self-awareness and an ability to navigate the world during waking hours.

The dreaming brain employs all of the same systems and networks, Dr. Hobson said, but with a few critical differences. Input from the outside world is screened out. Self-awareness ceases. The body is paralyzed. And everything that the dreaming brain sees, hears or feels is generated from within.

During sleep the non-dreaming brain falls physiologically somewhere between these two states, Dr. Hobson said. New technology allows researchers to examine the chemical, electrical and physical properties of each state. "We can study each of these states to see what the differences are," he said.

The key to dreams is found in several tiny nodes within the brain stem that contain cells which squirt out different chemical transmitters, Dr. Hobson said. These cells have projections that carry the chemicals throughout the brain and modulate its activity. The projections also extend down to the spinal cord and help control movement.

The awake brain is dominated by so-called adrenergic chemicals released from two of these nodes, Dr. Hobson said. The cells fire these chemicals in pacemaker fashion, keeping the brain alert, enhancing attention and priming motor activity. In the adrenergic state thought processes are generally stable and the brain is not easily dominated by stray images.

As the brain goes to sleep, the adrenergic system begins to shut down, Dr. Hobson said. Other nodes release so-called cholinergic chemicals which, although active in performing important brain functions during the day, begin to dominate the brain's chemistry. Self-awareness ceases and memory is lost. The brain enters a state of dreamless sleep featuring fuzzy images. It is not organized.

The cholinergic neurons are located one millimeter away from neurons that control muscles in the eyelids, Dr. Hobson said. This is why the eyelids get heavy when cholinergic signals take over. "It's the body's gimmick to get you to sleep," he said.

Soon, Dr. Hobson said, adrenergic neurons cease firing altogether. Cholinergic chemicals decouple the extensive networks used for cognition and behavior. In this state, the brain is ultrasensitive to stray thoughts and can jump from one class of images to another without realizing contradictions. In other words, it dreams.

The bifurcations of thought and the bizarre nature of dreams are also driven by cholinergic neurons, he said. Instead of firing steadily, as they do during the day, these cells begin bursting hundreds of times a second and sending erratic pulses into higher brain regions. These pulses, called PGO (or pontine-geniculate-occipital) waves, occur only during dreams and have multiple effects.

First, they stimulate the body's motor centers, found in the brain stem. This would cause the dreamer to walk, run and carry out a vast repertoire of movements except that another signal is sent simultaneously to the spinal cord, resulting in total muscle paralysis except for the eyes. This increase in cholinergic activity is what makes many people twitch or startle when they are drifting off to sleep, Dr. Hobson said. And it explains why sleepwalking almost always occurs during non-dream sleep, when the muscles are not paralyzed.

In dreams, Dr. Hobson continued, complex motor patterns are activated. "It's no accident that we are always in motion," he said. "We are practicing all sorts of movements in a kind of neural gymnastics."

PGO waves also bombard the brain's emotional circuits and give rise to the strong feelings that often accompany dreams, Dr. Hobson said. One third of all adult dreams involve anxiety and fear, he said, followed by joy, anger, sadness, guilt and eroticism.

Finally, the PGO waves shoot into the brain's higher regions where vast networks for processing information reside. Except now, Dr. Hobson said, every sight, sound and sensory input is generated internally by the brain itself. The higher networks, which are used to making representations of the world during the waking state, try to make sense of the internal images and feelings by concocting stories in the cholinergic brain, he said. The repertoire of possible combinations of images, memories and story lines on any given night is extremely large, and chance stimulation probably plays a large role in the content of our dreams.

People can enter a cholinergic state without going to sleep, said Dr. Edgar Garcia-Rill, a neuroscientist at the University of Arkansas in Little Rock, in an interview after the Boston meeting. He points out that such states exist during meditation, as well as sometimes when people are in sensory isolation tanks. Pilots occasionally experience these states when they stare at nothing but blue sky. Hallucinations, as in dreams, may result.

Many scientific insights have occurred in dreams, when the brain is open to unusual associations, Dr. Garcia-Rill said. He suggested that poets may be more naturally cholinergic. And, in work under way in his laboratory, Dr. Garcia-Rill has found that some schizophrenics have an abnormal number of cholinergic neurons in their brain stems, and that this defect may explain the hallucinations that mark the disorder.

To explore dreams in a more natural setting than the hospital-based sleep laboratory, Harvard researchers recently invented a device called the Nightcap. It is a cloth bandanna, worn pirate-style around the head, attached to a wallet-sized instrument that tucks under the pillow. One lead goes to a tiny box stitched into the bandanna that registers each time a person's head moves. A second lead sticks to one eyelid and records lid movements all night long. The lids move during dream sleep, whereas head movements tend to occur when dreams end and non-dream sleep or the transition to wakefulness occurs, said Robert Stickgold, a researcher in the laboratory.

The Nightcap can record up to 30 nights of sleep, Mr. Stickgold said. The device plugs into a computer so that any individual's sleep patterns can be analyzed in minutes, easily and relatively inexpensively, while he sleeps at home in his own bed. A Nightcap user can program the device to sound a signal during a dream episode to wake him up, so that he can recall the dreams.

Home dream reports are four times longer than dream reports taken in laboratories, Mr. Stickgold said.

In experiments described in the journal *Consciousness and Cognition,* the Nightcap showed that alcohol use suppresses dream sleep, that Prozac intensifies it and that heavy exercise can result in no body movements throughout the night.

Mr. Stickgold said that researchers in Thailand plan to use the device to explore why many young men who live near the Laotian border die suddenly in their sleep. The current explanation is that the men are being visited by the widow ghost who punishes them for infidelity.

"We also hope to send the Nightcap into space," Mr. Stickgold said. Astronauts on the Mir space station could use it to study the effects of zero gravity on dreams: "Do you still shift your body after coming out of a dream? Are dream reports going to be different?" he said.

The Harvard laboratory is also studying the content of dreams for insights into changing brain states. Dreams are full of discontinuities, incongruities, uncertainties and jumps in plot, scene location, characters and objects. What do these say about dream mentation?

Using 453 dream reports from 45 volunteers, the Harvard researchers cut dream reports into segments at discontinuity points—as when a beach scene abruptly and nonsensically turns into an indoor scene. The segments were spliced together in novel ways, while other dreams were left intact. Independent observers were asked to pick out the spliced dreams. They did no better than chance.

In a second experiment, the researchers studied character and object transformations in dreams. Can a teacher turn into a whale? Can a barn turn into a teacher? Apparently not, for when people were shown two columns of characters and objects and asked to pair those that made sense, they got it right 80 percent of the time.

"There are limits to the transformations that can occur at some deep category level," Dr. Hobson said.

Eventually, he added, research into sleep and dreaming may help explain the nature of consciousness. People cycle through different brain states every 24 hours, he said. If more can be learned about each state, it might be possible to subtract one from the other to learn how consciousness is engineered.

—SANDRA BLAKESLEE, July 1994

How the Brain Might Work: A New Theory of Consciousness

FOR SCIENTISTS who study the human brain, even its simplest act of perception is an event of astonishing intricacy.

Consider this: It is a beautiful spring day and you are walking down a country lane, absorbed in thought. Birds are chirping, roses are in bloom and the sun feels warm on your face. Suddenly, you hear a dog bark and you switch your attention to see if the animal means to bite.

Years of research have shown that the brain absorbs a scene like this by carving it into components and analyzing each chunk of information along separate pathways. As the eyes gaze at the rose, it is not the whole image of the rose that is transmitted to the brain. Instead, something very puzzling takes place. The nerve cells in the retina immediately break down the image into separate components, like its contours, textures and colors. As the ear hears birds chirping, separate cells respond to each frequency while others compute the direction and intensity of the sound. Cells in the skin that respond to warmth channel their input to yet another part of the brain.

Each population of sensory cells, from the eye, ear, nose and skin, sends its information to its home area on the outer surface of the brain, a thin, deeply furrowed sheet of cells known as the cerebral cortex.

The sensations of one instant of a spring morning have thus become represented by millions of activated cells in many different regions of the cortex. That much is known. A still baffling question for scientists is, how does the brain bind these fragmented pieces of information into a single coherent image? The nature of the reassembly process, known as the binding problem, is intimately related to the age-old question of consciousness, since an answer to the binding problem would go far to defining the physical basis of the conscious mind.

The first step to understanding, brain scientists say, is to realize that there is no Cinemascope screen in the brain where all the pieces come together. But if there is no screen, on what physical principle is consciousness organized? A growing number of scientists say the answer must lie in some form of timing. An image may be reconstructed from all cells that are firing in a particular rhythm at a particular instant.

Recent experiments have shown that precise timing codes are the brain's primary organizing principle, at least at the level of individual neurons, among specialized groups of neurons and across different parts of the brain.

But exactly how the timing codes work is a matter of vigorous debate. "Cells do carry information by virtue of the fact that they are firing at the same time," said Dr. Nancy Kopell, a biologist at Brandeis University in Waltham, Massachusetts, who studies how creatures move. "But what this means for function is unclear." Different solutions proposed for the binding problem have produced a lot of "yelling and screaming," Dr. Kopell said.

Nevertheless, efforts to understand how the brain uses time are forging ahead, said Dr. Christof Koch, a neuroscientist at the California Institute of Technology in Pasadena who is a leading theorist on the nature of human consciousness. The challenge is to construct theories "firmly based on nerve cells, their firing properties and their anatomical connections," he said. The brain may have evolved different binding solutions for different levels of organization.

The search for timing codes gets more speculative at the level of cell populations, Dr. Koch said. The basic idea is that cells involved in forming a perception will fire simultaneously, thus binding together in time rather than space. Every perception would be based on the temporary activation of an ensemble of neurons, he said. When a new perception is formed, the previous ensemble falls away and a new grouping of neurons fires, forming a new perception. Single neurons can participate in the representation of many things, depending on the ensembles they join in any one instant.

At the University of California at Davis, Dr. Charles Gray is recording the electrical activity of brain cells in different parts of the monkey visual system. There is a growing amount of evidence that cells fire in synchrony, he said. The problem is knowing if such synchrony is related to behavior—something no one has yet proved.

And even if cells fire synchronously, which cells are they? Is there something special about them? Dr. Gyorgy Buzsaki, a neuroscientist at Rutgers University in Newark, New Jersey, thinks there is. He has found that a class of cells called inhibitory interneurons have an inherent tendency to fire in a wave-like pattern. From the way they are distributed in the brain, these neurons could perform a binding function, he said.

"You can compare it to traffic control in New York City," Dr. Buzsaki said. "Say you have an imaging device that looks at 5,000 vehicles—cars, trucks, taxis, bicycles—all moving together in chunks. You'd like to figure out how they interact to achieve this togetherness. One answer is traffic lights," he said. Like traffic lights, interneurons are rhythmic and well coordinated and could control the flow of cognition. When interneurons are damaged, he said, the result is epilepsy—a storm of uncoordinated electrical activity in the brain.

Different brain regions may have evolved their own temporal binding codes, said Dr. Rodolfo Llinas, a professor of neuroscience at New York University. The motor system is a good example. The cerebellum is a structure that coordinates movements, relaying a barrage of signals from higher brain regions where decisions are made to the muscles. How is this coordination achieved?

The brain stem contains a nucleus of cells that burst at a rate of 10 cycles per second. These cells, inferior olivary neurons, send long fibers up to the cerebellum, where they make dense connections, thus amplifying their signals. Information flowing into the cerebellum is regulated by these bursting cells, making sure that movements only occur 10 times a second, Dr. Llinas said. The oscillation literally binds brain commands with muscle movements.

"This means we move in a non-continuous manner," Dr. Llinas said. "No one can move faster than 10 times a second because that is the normal frequency we all have. We have the impression of fluidity of motion because it all happens so fast." By binding packets of motor nerve information every tenth of a second, he said, the system has time to send messages to muscles more or less synchronously. The dynamic rhythm allows different combinations of muscles to be used for movements.

The same kind of binding mechanism may exist for the entire brain at a faster frequency of 40 cycles per second, Dr. Llinas said. At New York University he and a colleague, Dr. Urs Ribary, have measured such signals in human brains using a machine that detects magnetic fields on the scalp. The

40-cycle-per-second wave continuously sweeps the brain from front to back every 12.5 thousandth of a second, Dr. Llinas said, and could be the binding signal that links information from the parts of the cortex that handle auditory, visual, motor and other sensory signals.

Dr. Llinas believes that the 40-cycle-per-second wave serves to connect structures in the cortex, where advanced information processing occurs, and the thalamus, a lower brain region where complex relay and integrative functions are carried out. He, too, has a candidate population of cells—the intralaminar nucleus in the thalamus—that might generate the binding signal.

The intralaminar nucleus, a kind of doughnut of cells within the thalamus, is an intriguing structure. Its nerve cells send out long axons that reach to every part of the cerebral cortex. Significantly, there are also returning axons that come down from all areas of the cortex back to the intralaminar nucleus. The thalamus and the cortex are thus connected in a special way.

Intralaminar cells fire in a natural pattern of 40 cycles per second, Dr. Llinas said, and he believes it is their firing rhythm that is the source of the rhythmicity he detects at the surface of the cortex. His idea is that in each cycle a wave of nervous impulses radiates out from around the intralaminar nucleus to all parts of the cortex above, much like the central arm of an old-fashioned radar screen illuminates each object in its path.

The thalamus has another important connection with the cortex, which is that all the body's sensory systems send their inputs to relay stations of nerve cells and then to their own regions of the cortex. All these relay stations are located within the thalamus, which is able to influence their firing patterns. As visual signals come in from the eye, the thalamus insures that active cells in the visual cortex are coordinated into a rhythm of electrical activity that is at or near 40 cycles a second.

Critical to the act of cognition, in Dr. Llinas's theory, is the interaction between the two systems of electrical activity, the active sensory cells in the cortex and the scanning wave from the thalamus. The way the brain creates images, in his view, is as follows. The wave of impulses from around the thalamus's intralaminar nucleus polls all the sensory regions mapped out across the cerebral cortex once every 12.5 thousandth of a second. The regions that have active cells, representing some sensory input, are entrained to the same rhythm as the scanning wave, and send back a train of nervous impulses to the thalamus, all precisely timed in a coherent pattern.

All the coherent impulses that are received in a given cycle are perceived as a single image, Dr. Llinas suggests. The sensory messages of sight, sound, smell and touch, are thus bound together not in a single place but in a single instant of time, Dr. Llinas suggests. When the eye no longer sees an object, those cells no longer respond to the thalamus's scanning system. Each scan—12.5 thousandth of a second in length—creates a new image, but the images come so quickly that they seem continuous, as do the frames of a movie.

The brain has several natural oscillatory states, which correspond to deeply engrained functional states, Dr. Llinas said. For example, when thalamic cells reverberate at two cycles per second, the brain is in a state of deep sleep. At 10 cycles per second, the human brain is awake but not paying attention to the outside world. And at 40 cycles per second it is either wide awake or vividly dreaming.

In walking down the sunny path, absorbed in thought, a person would be generating regular 40-cycle-per-second rhythms while constructing internal images of the external world, according to Dr. Llinas's theory. As long as the internal images agree with what is happening outside, the brain keeps updating the scene with a steady rhythm.

But when the dog barks, posing a threat, the 40-cycle-per-second signal is abruptly reset so as to incorporate the novel stimulus into the overall scene so that the new information can be dealt with.

Dr. Llinas's theory, if true, would resolve the binding problem and also go far to explaining the nature of consciousness. Consciousness is the dialogue between the thalamus and the cerebral cortex. The brain is an organ, and its function is to create images. At night, these images are dreams; during wakefulness, the images are modulated by the senses and represent the outside world to which they correspond in some very practical way that has been determined by evolution. A person's waking life is a dream guided by the senses, Dr. Llinas said.

Dr. Llinas's theory rests on his measurements of various electrical rhythms in the brain, as well as observations such as that when the intralaminar nucleus is damaged, people fall into a deep coma. The critical 40-cycle-per-second rhythm he has measured at the surface of the cerebral cortex had not until recently been detected by others. Such waves could be noise rather than a mechanism for temporal binding, said Dr. Chris Wood, an expert on brain imaging at the Los Alamos National Laboratory in New Mexico.

But Dr. Llinas's theory has received strong support from recent experiments conducted by Dr. Mircea Steriade at the Laval University School of Medicine in Quebec. He and his colleagues implanted electrodes into interconnected areas of a cat's thalamus and cortex, such as the eye's relay station in the thalamus and in the visual cortex.

When the animals were wide awake or dreaming, Dr. Steriade said, cells in both areas would oscillate together at 40 cycles per second for just a brief period. The shared rhythm quickly appears and disappears, he said, which is why others have found it hard to detect.

Dr. Steriade said that in his view the 40-cycle-per-second rhythms "exist spontaneously when animals are in active states of vigilance."

Other neuroscientists are withholding judgment on theories like Dr. Llinas's until the various rhythms or oscillations of nerve cells in the brain are better understood. "There is no question that oscillations exist," said Dr. David Hubel, a leading expert on vision at Harvard Medical School. "But we have no idea what, if anything, they are doing" in the human brain. One can think about temporal coding, he said, "but it's very hard to drum up experiments that get at the problem.

"One has to grant that of the 200 possible cortical areas in the brain, very few have been explored in detail," Dr. Hubel explained. Thus there could be as yet undiscovered, highly specific areas in the brain where information comes together and is bound into a unitary experience.

"No one in neuroscience thinks time is not important," said Dr. Patricia Churchland, a philosopher and brain scientist at the University of California at San Diego. "Criticisms arise with how time management is achieved." Theories like those proposed by Dr. Llinas are an excellent entry into the binding problem, she said, but "do not yet solve the problem."

—SANDRA BLAKESLEE, March 1995

The Conscious Mind Is Still Baffling to Experts of All Stripes

LIKE THE PROVERBIAL blind men trying to identify by sense of touch a large, thick-hided animal with a trunk at one end and a tail at the other, some the world's top scientists, philosophers and far-out thinkers gathered here last week to contribute their different perspectives on the elephant of consciousness.

A good time was had by all, even when the fur—or maybe it was elephant hide—began to fly.

Can machines be conscious? The question elicited a spirited debate between those who said, Of course, it's just a matter of time and clever engineering, and others who replied: Never! It's bad enough that you think consciousness can arise from gray lumps of tissue. It is inconceivable that sentience could ever emerge from wholly insentient matter.

Then there were less contentious questions. Does free will exist? Can consciousness exist without emotions? Are animals conscious? What happens to your conscious mind when you fall into a deep sleep?

And the most debated question of all: Is consciousness something very special and unique or is it just the natural byproduct of a complex brain, emerging like wind from intricate weather patterns?

The conference, "Toward a Science of Consciousness," which attracted 1,000 participants from 22 countries, was sponsored by the University of Arizona with support from the Fetzer Institute and the Institute of Noetic Sciences, two organizations dedicated to exploring the metaphysical foundations of Western science. A similar but smaller conference was held two years ago in Tucson.

Thus the conference drew neuroscientists, philosophers, mathematicians, computer scientists, physicists, dream researchers, pharmacologists, doctors, ethnologists, psychologists, parapsychologists, scholars of religion and a variety of prophets who claim to have solved the mystery of consciousness.

The meeting was unusual from the start. Dr. Jaron Lanier, a computer scientist from Columbia University who is a pioneer in virtual reality, opened the plenary session on Monday by playing a brief piano recital. His blond dredlocks flew apace with the music. The audience was delighted.

The goal of the meeting was simple, said Dr. Stuart Hameroff, an anesthesiologist at the Arizona Health Sciences Center who was a principal organizer of the event. What is the nature of consciousness? Can we hope to understand it scientifically?

It is remarkable that such a diverse gathering could discuss the question of consciousness in a coherent manner. But this kind of cross pollination of ideas, where everything goes, is exactly what is needed, said Dr. Christof Koch, a neuroscientist at the California Institute of Technology who helped organize the event. One hundred years ago, people could not understand how life could arise out of mere chemicals, he said. But when DNA was explained, theories of vitalism—that a magical force was needed to explain life—disappeared.

The study of consciousness is like the study of physics before Newton, said Dr. Piet Hut, a theoretical physicist at the Institute for Advanced Studies in Princeton, New Jersey. In fact, he said, if people had organized a conference about physics in the Middle Ages, they would have dismissed Copernicus and Galileo as crackpots. "We shouldn't make that mistake today," Dr. Hut said.

But before progress can be made on the question, some definitions are in order. Consciousness has many guises.

In Tucson, the tone of discourse was set by a young philosopher from the University of California at Santa Cruz, Dr. David Chalmers. He is widely credited for posing the so-called hard problem of consciousness.

To explain this concept, Dr. Chalmers first described the so-called easy problems of consciousness, the sorts of questions being tackled in neuroscience laboratories around the world: How does sensory information get integrated in the brain? How do we see and reach out for an object? How are we able to verbalize our internal states and report what we are doing or feeling?

"These problems are not trivial," Dr. Chalmers said. "They may take 100 years or more to solve, but progress is being made."

The hard problem is this: What is the nature of subjective experience? Why do we have vividly felt experiences of the world? Why is there someone home inside our heads?

Thus far, nothing in physics or chemistry or biology can explain these subjective feelings, Dr. Chalmers said. "What really happens when you see the deep red of a sunset or hear the haunting sound of a distant oboe, feel the agony of intense pain, the sparkle of happiness or meditative quality of a moment lost in thought?" he asked. "It is these phenomena, often called qualia, that pose the deep mystery of consciousness."

In Tucson, people mounted four responses to the hard problem: It doesn't exist, it will be answered soon enough by conventional science, there must be something else in the universe that we do not yet understand, and hey guys, forget it, we can never understand consciousness.

Dr. Daniel Dennett, a philosopher at Tufts University, is a forceful proponent of the idea that consciousness is no big deal. "It's like fame," Dr. Dennett said. "It doesn't exist except in the eye of the beholder." When does fame happen, he asked? Is it when 10 people know your name? A hundred people? A thousand?

Scientists have shown that information coming into the brain is broken down into separate processing streams, Dr. Dennett said. But no one has yet found any "place" where all the information comes together, presenting a whole picture of what is being felt or seen or experienced. The temptation, he said, is to believe that the information is transduced by consciousness. But it is entirely possible that the brain's networks can assume all the roles of an inner boss. Mental contents become conscious by winning a competition against other mental contents, Dr. Dennett said. No more is needed. Consciousness is an epiphenomenon.

A second group of scientists agreed with Dr. Dennett but took a softer line. When all the "easy" problems are solved, the hard problem will disappear—but consciousness certainly exists. "It's silly to deny it," said Dr. Patricia Churchland, a philosopher at the University of California at San Diego.

Awareness and subjectivity are network effects involving many millions of nerve cells in the cortex and thalamus, Dr. Churchland said. And while the exact nature of the phenomenon cannot yet be explained, the call for a "new physics" or some mysterious forces in nature are not needed.

Neuroscientist Dr. Rodolfo Llinas of New York University agreed, suggesting that timing effects inside the brain produce conscious experience.

Those who believe machines can someday be conscious tended to fall into this camp. The trick will be to make computers that are sufficiently complex, said Dr. Danny Hillis, vice president of research and development at Walt Disney Imagineering in Glendale, California. Then, like human brains, they should give rise to the emergent properties of consciousness.

Others tried to answer a few of the easy questions. Dr. Allan Hobson, a sleep expert at Harvard Medical School, described a neurobiological theory of dreaming. It does not explain where consciousness "goes" when people are asleep, he said, but finds that different chemical states in the brain seem to produce different sorts of consciousness.

The next major group of consciousness seekers might be call modern dualists. Agreeing with the hard problem, they feel that something else is needed to explain people's subjective experiences. And they have lots of ideas about what this might be.

According to Dr. Chalmers, scientists need to come up with new fundamental laws of nature. Physicists postulate that certain properties—gravity, space-time, electromagnetism—are basic to any understanding of the universe, he said. "My approach is to think of conscious experience itself as a fundamental property of the universe," he said. Thus the world has two kinds of information, one physical, one experiential. The challenge is to make theoretical connections between physical processes and conscious experience, Dr. Chalmers said.

Another form of dualism involves the mysteries of quantum mechanics. Dr. Roger Penrose from the University of Oxford in England argued that consciousness is the link between the quantum world, in which a single object can exist in two places at the same time, and the so-called classical world of familiar objects where this cannot happen. Moreover, with Dr. Hameroff, he has proposed a theory that the switch from quantum to classical states occurs inside certain proteins call microtubules. The brain's microtubules, they argue, are ideally situated to perform this transformation, producing "occasions of experience" that with the flow of time give rise to stream of consciousness thought.

The notion came under vigorous attack. "Pixie dust in the synapses is about as explanatorily powerful as quantum mechanics in the microtubules," Dr. Churchland said. Their logic is, consciousness is deeply mysterious, quantum mechanics is deeply mysterious, ergo the two are the same mystery, she said.

Dr. Penrose's ideas are popular, Dr. Churchland said, because many people have dualist hankerings. They want to believe in a soul, life after death and the specialness of humans and their inner thoughts. They have a negative gut reaction to the idea that neurons—cells that can be probed under a microscope—are the source of the "me-ness of me," she said.

Finally, there are those who argued that people can never understand consciousness. The mystery is too deep. Dr. Colin McGinn, a philosopher from Rutgers University, said that for humans to grasp how subjective experience arises from matter "is like slugs trying to do Freudian psychoanalysis—they just don't have the conceptual equipment."

But this did not deter many from trying. During the week, presentations were made on animal consciousness (featuring apes, dolphins and gray parrots), free will and the spiritual nature of consciousness.

Dr. Robert Forman, a professor of religion at Hunter College in New York, said mystical experience had something to tell people about consciousness. "To understand genes," he said, "we look at bacteria like *E. coli.* To study memory, we analyze the memory of a sea slug. But to probe consciousness, we need to examine the experience of mystics, who experience their own consciousness in its simplest form."

Millions of people regard these types of experiences, feeling a oneness with the universe, as the highest experience that the conscious brain has to offer, Dr. Forman said.

—Sandra Blakeslee, April 1996

9

THE LATEST FROM THE FIELD

Two remarkable advances in brain science have been made since the articles in this book were first compiled.

One is the creation of a strain of genetically smarter mice. The other is the dawning realization that the brain's complement of cells is not fixed at birth but that new cells are generated continually throughout life.

Smarter mice may not be of any great use in themselves. They were generated, by Dr. Joe Tsien of Princeton University, to prove an idea about how memory traces are laid down at the level of the individual brain cell or neuron. Building on the work of many other scientists, Dr. Tsien figured out a critical component in the mechanism of memory formation, along with the genes whose proteins make up the mechanism's parts. By making one of the genes more active than usual in the mice's brain cells, he succeeded in generating a strain of mice with improved memory. The enhanced memory translated into better performance on a standard set of intelligence tests designed for laboratory mice, confirming that good memory is part of intelligence.

A first reaction to Dr. Tsien's work is that maybe human intelligence can be enhanced, whether by genetic engineering or, more probably, by drugs that might mimic the effect of the activated gene. Such drugs should certainly benefit people whose intelligence was impaired by disease or injury. But those eager to raise their IQ from normal to prodigy levels might pause to ask the very interesting question of why nature, in her wisdom, has neglected to maximize the mouse's intelligence. Could it be that with too good a memory, a mouse would be overly fearful? Does forgetfulness contribute as much as omniscience to survival? And if evolution has made mice exactly as smart as they need to be and not a whisker more, perhaps human societies work best with members of our limited cognitive capacities.

Be that as it may, Dr. Tsien's smart mice demonstrate the penetrating level of understanding that neuroscientists have now achieved into how the neuron performs one of its most basic housekeeping tasks.

Another significant advance has emerged from studying the dynamics of the brain's cells. Scientists have long believed that the number of neurons in the human brain—some 100 billion—is fixed by the time of birth and that no new cells are added. This dogma has gradually been overturned. Biologists now know that the brain, like other renewable tissues such as the skin and stomach lining, possesses its own cache of stem cells. These stem cells, unlike mature neurons, can grow and divide in profusion, creating a stream of new neurons.

The brain consists of many different regions or modules. At least two of these modules receive regular reinforcements of new neurons from the stem cells. One is the olfactory bulb, which mediates the sense of smell. The other is the hippocampus, where new memories of faces, places and maybe other things are initially laid down.

An intriguing study by Elizabeth Gould and Charles Gross of Princeton, not yet fully accepted by other biologists, holds that the cerebral cortex too receives a stream of new neurons. The cerebral cortex, the outer rind of the brain, is the seat of cognition and higher brain functions. A particularly intriguing suggestion is that this regular stream of new cells is somehow the basis of historical memory. After all, what better way of storing memories than to create a cohort of new cells each day, stored in order of their birth dates, and make each cohort the repository of that day's memories. No wonder that memories fade with age: like the bottom layers of ice in the Greenland ice cap, the oldest layers get thinned out almost to nothing by the pressure of all the younger layers above, so that only cells recording the most evocative memories are retained.

This elegant idea is so far only just that; other scientists have not yet confirmed the Gould-Gross findings. Still, it is clear that the brain is a hotspot of new neuron generation throughout adult life.

Many biologists hope that the brain's stem cells can be used to repair the cells lost in certain diseases, such as the dopamine-producing cells whose loss causes Parkinson's.

The brain is dynamic not only in the sense of being able to create new cells but also, at least in certain stages of life, of being able to rewire itself. If the inputs to the brain from baby ferrets' eyes and ears are switched around, the animals learn to see with their auditory cortex and to hear with their visual cortex. This amazing plasticity shows how much the organization of the infant brain depends on the input to it. Some special inhibitory mechanism seems necessary to prevent the different senses interfering with one another. In some people, known as synesthetes, the inhibition seems to be incomplete, and colors may be strongly associated with sounds. Drugs like LSD and mescaline also produce this commingling of senses, as if by reversing the normal inhibition.

The plasticity of the infant brain and synesthesia are two examples of how much remains to be understood about the brain above the level of its genes and cells.

—NICHOLAS WADE

Scientist at Work: Joe Z. Tsien; Of Smart Mice and an Even Smarter Man

A CERTAIN AMOUNT OF DISORDER has broken out around Dr. Joe Z. Tsien, the biologist who announced last week that he had created a smarter strain of mouse by genetically altering a gene for memory. Patients call seeking help. Individuals of enhanced imaginations warn that the mice may escape and take over the planet. Television crews patrol the halls. His voice-mail box has overflowed.

But Dr. Tsien, seemingly the only scientist on the Princeton campus who, on a warm summer day, is wearing a tie, ignores the chaos and a phone that rings every couple of minutes. In soft tones he describes the remarkable journey that has led him from Wuxi, a small town near Shanghai, to the position of having made a significant, maybe decisive, contribution to understanding the nature of memory and intelligence.

Dr. Tsien (pronounced chee-YEN) says he did not begin to consider the wider implications of his work until just before his article was published. He engineered his smarter mice for purely academic reasons, to address and perhaps solve the question of how memories are laid down in the brain. But the mice turned out to be smarter as well as having better memories, lending an unexpected new dimension to the experiment.

Although many arguments with psychologists doubtless lie ahead, Dr. Tsien believes that learning, memory and intelligence are all intimately related because, as his smarter mice demonstrate, "a common unifying mechanism underlies them all."

And because mice and people use the same basic mechanism of memory, the smarter mice could well shed much light on the nature of human intelligence.

Dr. Tsien's result, as he is the first to note, rests on knowledge and techniques developed by other scientists. He describes his experiment as "obvious"—at least in retrospect. His achievement lies in the fact that, in a highly competitive field of biology, he was the first to conceive of the experiment and to see that it could be decisive. He also carried it out in a particularly convincing way. "Extremely nicely done," was the verdict of Dr. Eric R. Kandel, a leading biologist at Columbia University and the former laboratory chief of Dr. Tsien.

The idea that led to the smarter mice was no lucky break. Rather, it was a feat for which Dr. Tsien had been preparing intensively for many years, including seven years of postdoctoral education. In Wuxi, where his father was a clerk and his mother an accountant, he was the only person to enter college from his high school, one attached to a fabric plant. But the college was a good one, the East China Normal University in Shanghai, and he decided to do doctoral studies in the United States.

"In 1986, China was still very closed, so we really had no idea about the United States," Dr. Tsien says in describing how he picked a college. He chose the University of Minnesota because it offered to waive the application fee, which he could not afford, and because the Chinese characters for Minnesota translated invitingly to "clean air blue sky."

Having recovered from the surprise of finding the clean-air-blue-sky state so cold, he developed an interest in neurophysiology and the instruments then available for monitoring the electrical signals transmitted by brain cells. "I got fascinated by seeing a nerve cell fire. They are talking—what does that mean?" he says.

A long apprenticeship was necessary before he could begin to parse that language. He did his Ph.D. thesis with Dr. Lester R. Drewes of the University of Minnesota, helping him conduct studies under a Defense Department grant on how the warfare agent sarin blocks the transmission of nervous signals.

Receiving his Ph.D. in 1990, he was accepted as a postdoctoral student by Dr. Kandel's laboratory. There he worked on identifying genes that are active in rats' brains during memory formation. "I got a more systematic education in neuroscience. I got to see how a big lab operates," Dr. Tsien said.

He then moved to another leading neuroscience laboratory, that of Dr. Susumu Tonegawa at the Massachusetts Institute of Technology. Dr.

Tonegawa won the Nobel Prize in Physiology or Medicine in 1987 for research on the genetic control of the immune system, and later switched to the study of learning.

In Dr. Tonegawa's lab, Dr. Tsien worked with so-called knock-out mice, animals from which a gene has been deleted. The idea is to learn what a gene does by excising it and seeing what defects the mouse develops. He became interested in the brain cell component, known as the NMDA receptor, suspected of being central to the memory mechanism. The receptor consists of parts made by several genes, the chief part being specified by a gene called NR1.

Using advanced genetic techniques, he decided to create a mouse lacking the NR1 gene in the cells of its forebrain. Creating the mouse took two and a half years and, for a postdoctoral student, was a substantial risk. If the experiment failed, there would be no result worth publishing.

In the end, he was able to knock out the gene in just the cells of the hippocampus, a brain module dedicated to learning and much studied by neuroscientists. "I think a god looked on me very kindly," he said, referring to the element of luck in creating such a valuable research tool.

The mice lacking the NR1 gene in the hippocampus indeed did not remember as well, suggesting the NMDA receptor is important in laying down memories. But the experiment, published in December 1996, was regarded by other experts as less than fully conclusive, because the absence of the NR1 gene could have caused general brain damage not specific to memory.

Dr. Tonegawa became very interested in the mouse, as did Dr. Kandel, because Dr. Tsien had made it with a technique developed in Dr. Kandel's laboratory. The two lab chiefs were also interested in receiving due credit, and discussions ensued between them that were stressful for him, Dr. Tsien recalls.

However, he now had sufficient credentials to set up his own laboratory. "After working with these two powerful people, I wanted to be free," Dr. Tsien says. Two years ago he was appointed an assistant professor at Princeton and was able to set up his own lab. He began to think about how he might try to improve a mouse's memory, rather than sabotage it, because such an experiment would run far less risk of being criticized as nonspecific.

The focus of his thinking was the anatomy of the NMDA receptor. The intricate biological device is shaped like a cylinder or doughnut embedded in the outer wall of certain brain cells. Usually its central channel is firmly closed. But when the nerve cell receives signals from two other nerve cells at the same time, the NMDA channel springs open, allowing a current to flow into the cell. This current generates a long-lasting change within the cell, making it much more responsive the next time that either of the two other nerve cells is active alone.

This property of the NMDA receptor—opening when two signals arrive simultaneously—has long been suspected to be the basic mechanism of memory, because it is a way for the brain to make an association between two events. The exact degree of simultaneity turns out to be very important. In young mice, two signals can arrive as much as one-tenth of a second apart for their coincidence to change the nerve cell. In older animals, the NMDA receptor allows a much narrower window of time for an association to register.

Another known fact about the receptor was that its composition changed with the age of the animal. Its main component is the gene product of NR1, which Dr. Tsien had knocked out in his memory-deficient mice. But the NR1 component works with any of four different partners, which modulate its activity in different ways. Two of the partners, known as NR2A and NR2B, are particularly important in cells of the forebrain. As animals age, there is a switch from NR2B to NR2A as the preferred partner for NR1.

Abilities in many animals decline after sexual maturity. Song birds cannot learn new songs. The human mind becomes less flexible at learning new languages. "I am always stuck with my Chinese accent, but if I had come to the United States 12 years earlier I would have learned perfect American," Dr. Tsien says by way of personal example.

As he contemplated these various pieces of information, Dr. Tsien said, it seemed clear that they were related. The natural switch with age of NR1's partner must underlie the increasingly stringent requirement for two signals to arrive simultaneously, and the narrowed window of time must be the reason why older people find it harder to make associations.

But no one had specifically stated the idea in those terms, as far as Dr. Tsien knew. And certainly no one had done the obvious experiment, which was to engineer mice in which the NR2B gene was artificially put into hyperdrive to see if their memories improved.

So Dr. Tsien took a copy of the mouse NR2B gene and linked it to a special piece of DNA, called a promoter, that is active only in cells of the mouse forebrain. He injected this genetic fragment into fertilized mouse eggs, where it added itself to the mouse's normal complement of genes. Because of the promoter, the NR2B gene was active in cells of the forebrain, adding its product to that produced by the mouse's own NR2B gene.

With all the extra NR2B being produced in the mice's brain cells, the NMDA receptors underwent a subtle but significant change. Instead of staying open for 100 thousandths of a second, as they do in normal mice, the receptor's interval increased to 250 thousandths of a second.

That minute biophysical change, Dr. Tsien says, is what underlies the superior learning skills of the mice. The essence of smartness is an extra 150 thousandths of a second.

Credit for a discovery is often disputed, particularly when the finding is important. Dr. Tonegawa told reporter for *The Star-Ledger* of Newark, that Dr. Tsien may have started developing the smarter mice while at the Massachusetts Institute of Technology and accused him of being uncollegial. If the mice were developed in Dr. Tonegawa's lab, M.I.T. would have rights to them and Dr. Tonegawa could exercise a lab chief's claim to a share of the academic credit.

Dr. Tsien said he was "totally surprised" by Dr. Tonegawa's remarks. His smarter mice experiment was conceived and executed entirely at Princeton, he said. Other scientists have patented the NR2B gene but Princeton has filed for a "use patent," the right to use the gene in ways suggested by Dr. Tsien's work.

Dr. Tonegawa, who was traveling in Japan last week, did not respond to a request made through his secretary for an interview.

Dr. J. David Litster, M.I.T.'s vice president for research and the official in charge of disputes over scientific conduct, said that M.I.T. "is not endorsing Tonegawa's claims, certainly not until we know what they are. If it's a dispute between Tonegawa and a former postdoc over credit I don't really think we ought to get involved in that."

Meanwhile Dr. Tsien is waiting to see how the implications of his smarter mice are received by his peers in the neuroscience community and by the public. "To the scientific community this is a small step for a man," he says. "The fundamental question is 'Is this a giant step for mankind?' "

What does he think? "I don't know," Dr. Tsien replies.

—NICHOLAS WADE, September 1999

"Rewired" Ferrets Overturn Theories of Brain Growth

LIKE INVENTIVE ELECTRICIANS rewiring a house, scientists at the Massachusetts Institute of Technology have reconfigured newborn ferret brains so that the animals' eyes are hooked up to brain regions where hearing normally develops.

The surprising result is that the ferrets develop fully functioning visual pathways in the auditory portions of their brains. In other words, they see the world with brain tissue that was only thought capable of hearing sounds.

The findings, reported by Dr. Mriganka Sur and his colleagues in the April 20 issue of *Nature* magazine, contradict popular theories on how animal brains develop specialized regions for seeing, hearing, sensing touch and, in humans, generating language and emotional states.

Many scientists claim that genes operating before birth create these specialized regions or modules, arguing for example that the visual cortex is destined to process vision and little else. But the ferret experiments show that brain regions are not set in stone at birth. Rather, they develop specialized functions based on the kind of information flowing into them after birth.

"Some scientists are going to have a hard time believing these experiments," said Dr. Jon Kaas, a professor of psychology at Vanderbilt University in Nashville. They demonstrate, Dr. Kaas said, "that the cortex can develop in all sorts of directions."

"It's just waiting for signals from the environment and will wire itself according to the input it gets," he said. The findings may shed light on unusual brain patterns observed in people who are born deaf or blind, he added.

"If you wanted to create a dream experiment, this would be it," said Dr. Michael Merzenich, a neuroscientist at the University of California at San Francisco and a leading authority on the brain's ability to change and reorganize, a process known as plasticity. "It's about the most compelling demonstration you could have that experience shapes the brain."

The researchers are all members or former members of the department of brain and cognitive sciences at M.I.T. The rewiring experiments began more than 10 years ago, Dr. Sur said. He chose ferrets because their brains are very immature at birth and undergo a late form of development that the researchers can exploit. As in humans, the ferret's optic and auditory nerves travel through a way station called the thalamus before reaching areas in the higher brain or cortex where vision and hearing are perceived.

In humans, this very basic wiring is present at birth, but in ferrets, these important nerves grow into the thalamus after the animal is born. Dr. Sur found that if he stopped the auditory nerve from entering the thalamus, the optic nerve would arrive a few days later and make a double connection. It would go on through the thalamus and connect itself up to both seeing and hearing regions of the cortex.

The researchers then waited to see what would happen to the hearing region of the brain once it was getting all its signals from the retina.

After a ferret or human is born, cells in the brain's primary visual area become highly specialized for analyzing the orientation of lines found in images or shapes. Some cells fire only in response to vertical lines. If presented with a horizontal or slanted line, they don't do anything.

Other cells fire exclusively when a horizontal line falls on them and yet others fire in response to lines slanted at various angles. These specialized cells are draped across the primary visual area in a somewhat splotchy fashion that resembles a bunch of pinwheels.

The hearing region of the brain is organized very differently, Dr. Sur said. Each cell is connected to the next in a kind of single line. There are no pinwheel shapes.

After the rewired ferrets matured, researchers looked at the auditory region of their brains and found that cells were organized pinwheel fashion. They found horizontal connections between cells responding to similar orientations. The rewired map was less orderly than the maps found in normal visual cortex, Dr. Sur said, but looked as if it might be functional.

The researchers then asked, What does the rewired ferret experience? Does it see or does it hear with its auditory cortex?

Rewired ferrets were trained to turn their heads one way if they heard a sound and in the other direction if they saw a flash of light. In these experiments, one hemisphere was rewired and the other was left normal as a control. Thus the animals could always hear with the intact side of their brains and were deaf in the rewired side.

Not surprisingly, when the light was presented to the rewired side, the animals responded correctly. But when connections to visual areas were severed on the rewired side, the animals still responded to the light. It meant that they were seeing lights with their rewired auditory cortex, Dr. Sur said.

The research reopens the question of what are the relative contributions of genes and experience in building brain structure, according to Dr. Kaas.

Genes, Dr. Kaas suggests, create a basic scaffold but not much structure. Thus, in a normal human brain, the optic nerve is an inborn scaffold connected to the primary visual area. But it is only after images pour into this area from the outside world that it becomes the seeing part of the brain. All the newborn cortex knows about the outside world is from the electrical activity of these inputs, or images that fall on the retina, sounds that reach the inner ear or touch sensations that press on the skin, Dr. Kaas said.

As the inputs arrive, the cells organize themselves into circuits and functional regions. As these circuits grow larger and more complex, Dr. Kaas said, they become less malleable and, probably with the help of changes in neurochemistry, become stabilized. This is why a mature brain is less able to recover from injury than a very young brain.

Young brains are astonishingly plastic, Dr. Kaas said. For example, he said, children who suffer from a severe form of epilepsy that is treatable only by removing one-half of their brains can learn to walk, talk, throw balls and otherwise develop normally with only half a brain, if operated on early in life, he said.

But in recent years, scientists are also discovering that adult brains, as well, can undergo surprising changes in response to experience. For example, imaging experiments carried out on blind people show that when they learn to read Braille, "visual" areas of their brains light up. Touch seems to be residing in visual areas. Similar experiments on deaf people show that they use the auditory cortex to read sign language, whereas people who can hear use the visual areas of the brain for this purpose.

Dr. Sur said his laboratory was now searching for molecules that help produce these kinds of changes in mature and developing brains. If the chemistry of regrowth and reorganization can be understood, he said, it would offer new avenues for helping people recover from damage caused by strokes, accidents and various brain diseases.

—SANDRA BLAKESLEE, April 2000

When People See a Sound and Hear a Color

MOST PEOPLE, when not under the influence of hallucinogenic drugs, experience the sensory world as a place of orderly segregation. Sight, sound, smell, taste and touch live in different neighborhoods and commute on separate freeways: A Beethoven symphony is not pink and azure; the name Angela does not taste like creamed spinach.

Yet there are those for whom these basic rules of the sensorium do not seem to apply. They have a rare condition called synesthesia, in which the customary boundaries between the senses appear to break down, sight mingling with sound, or taste with touch.

Thus, the composer Olivier Messiaen, speaking of the union of color and tone in his music, explained to an interviewer: "When I hear music, I see inwardly, in the mind's eye, colors which move with the music. This is not imagination, nor is it a psychic phenomenon. It is an inward reality." A 21-year-old woman, participating in a continuing synesthesia study at the National Institute of Mental Health, told researchers that when she ate buttered toast, "it is rough, but not pointy; and if it has jelly on it the rough texture is rounded." And Carol Steen, a New York artist who, like most synesthetes, has had synesthetic experiences from an early age and who uses her perceptions in her work, says she distinguishes different types of headaches by their colors. "If it's a sinus headache, it's green," Ms. Steen said.

Synesthesia received a flurry of attention from artists and psychologists at the turn of the century. But until relatively recently, modern science largely ignored it. Those who experienced synesthesia rarely complained ("It's the most wonderful thing in the world!" exclaimed one synesthetic woman). And the private nature of the perceptions made investigation difficult—there was no objective way to tell what, if anything, unusual was taking place.

In the past 10 years, however, the arrival of imaging techniques and other new technologies for studying the brain at work has revived interest in synesthesia, capturing the interest of a small core of researchers in a variety of countries and disciplines. PET scanners, electrophysiological recording, DNA analysis and other techniques are increasingly being used. In the current issue of *The Journal of Neuropsychiatry and Clinical Neurosciences*, for example, German researchers from the University of Hanover Medical School report electrophysiological findings from a group of synesthetic subjects.

An understanding of synesthesia as a perceptual anomaly, researchers hope, may eventually help elucidate normal perception, or even shed light on consciousness itself.

Meanwhile, much more remains unknown about the comingling of the senses than is known. Even basic facts about synesthesia—its prevalence, for example—are still less than certain. One newspaper and magazine survey, however, by Dr. Simon Baron-Cohen, a psychologist, and his research group at Cambridge University in England, found that 1 out of 2,000 people reported synesthetic experiences.

Scientists offer differing theories of synesthesia's cause: Some argue that it represents an innate difference in neurophysiology, others that it is a result of associations learned at an early age. And researchers disagree about how exactly the condition should be defined, with some viewing it as a perceptual framework entirely distinct from normal perception, while others envision a continuum of synesthetic experience, with normal sensory perception at one end. Dr. Lawrence Marks, a psychologist at Yale University, for instance, has found that normal subjects show "implicit" associations that may be milder versions of the links found in synesthesia: Most people, for instance, associate brighter colors with higher pitched sounds.

Also mysterious is why the extent and form of synesthetic perceptions differ so widely from person to person. Seeing letters of the alphabet in different colors is a common synesthetic phenomenon. But, Dr. Baron-Cohen said: "Even within families, people argue about the colors of different letters. It seems to be highly individual and idiosyncratic."

And while the majority of synesthetes, who are about six times as likely to be female as male, find their unusual sensory abilities enjoyable, in some cases synesthesia can be disruptive. One woman, Dr. Baron-Cohen said, not only saw colors when she heard sounds, but heard sounds whenever she saw colors. "For her it was very unpleasant, and her reaction was to try to control the environment and keep everything very low key," he said.

Amid all this uncertainty, a growing body of evidence supports the notion that whatever synesthesia is, it represents more than the clever use of metaphor by creative individuals. Hallucinogenic drugs like mescaline and LSD, as well as some organic diseases, can produce synesthesia, suggesting a physiological basis. And people who report synesthetic experiences—seeing particular colors evoked by particular sounds, for example—demonstrate a remarkable consistency in their associations over time. In a 1989 study, Dr. Baron-Cohen and his colleagues found that synesthetic subjects showed a 92 percent consistency in their color-sound associations after a full year, in contrast to only 37 percent consistency after one week in control subjects.

Synesthesia also appears to run in families, leading some researchers to believe it has a genetic basis. Certainly, many synesthetes report that a family member shared their ability. The writer Vladimir Nabokov, for example, wrote that as a young child, he informed his mother that the painted colors on his wooden alphabet blocks were "all wrong." She understood immediately, Nabokov recalled, because she, too, saw each letter in a distinctive hue.

For her part, Ms. Steen remembers a family visit 30 years ago, when she was in college: Sitting at the dinner table, she mentioned casually that the numeral "5" was yellow, but her father corrected her. "No, it's yellow-ocher," he said. Her mother and brother, Ms. Steen said, sat in silent perplexity.

Perhaps most intriguing as arguments for the "realness" of synesthetic phenomena, however, are the handful of brain studies that show differences in brain functioning of subjects who have synesthesia. In 1982, Dr. Richard Cytowic, a Washington neurologist and the author of *The Man Who Tasted Shapes* (Putnam, 1993), led the way by measuring brain metabolism in a single synesthetic subject—a man who at a dinner party announced that the meal would be late because "there are not enough points on the chicken." Dr. Cytowic found that during synesthetic experiences, the man showed decreased blood flow in brain areas of the cortex responsible for language and abstract thought—findings the neurologist argued were indications that synesthetes were not simply using their imaginations or playing with language.

A more sophisticated PET scanning study, published in 1996 in the journal *Brain* by Dr. Eraldo Paulesu, Dr. Baron-Cohen and colleagues, compared brain functioning in six synesthetes to that in six members of a control group. The subjects, all women, were blindfolded and listened to sound cues delivered through headphones. Synesthetes, the researchers found, showed increased activation in some areas of the visual cortex when responding to sounds; control subjects did not.

The areas showing increased activity are not the same as those activated when someone is imagining images, said Dr. Baron-Cohen. He said an M.R.I. study, as yet unpublished, by researchers at the Institute of Psychiatry in London had replicated these results.

Most recently, the team of German researchers compared recordings of electrophysiological activity in the brains of 17 synesthetic subjects, who experienced color images upon reading letters or numbers, with recordings from 17 control subjects.

The results, said Dr. Thomas Muente, a neurology professor at Hanover who took part in the study, suggest that during synesthetic experiences, regions of the frontal cortex of the brain that control attention and also play a role in processing sensory information are inhibited. "But this paper is just the starting point," said Dr. Muente.

Murkiness in science is not without its rewards, and one benefit of the absence of conclusive data about synesthesia is that it has left ample room for the construction of titillating theories. Dr. Cytowic, for example, argues on the basis of his own research and the subsequent brain study findings that synesthesia is a kind of "cognitive fossil," left over from a time before the separation of sensory pathways evolved. In this sense, synesthesia may be "closer to the essence of what it is to perceive," Dr. Cytowic has said, and may be tied to phylogenetically older, subcortical brain structures involved in emotion and other primal functions.

Dr. Peter Grossenbacher, a senior staff fellow at the National Institute of Mental Health, who with Christopher T. Lovelace and Carol Crane has interviewed more than two dozen synesthetic subjects as part of a current study, has a different theory. He suggests that in synesthetic experiences, neural pathways that normally act to suppress irrelevant sensory input and allow focused perception in a single sensory mode, may be selectively disinhibited, resulting in multimodal perception.

For Patricia Duffy, a 46-year-old instructor in the United Nations' language and communication training program, the cause of her perceptions is less important than the richness they have brought to her life. She sees the words she speaks fly by in a rainbow of colors. She sees a year as an oblong circle, a week as a sidewalk with seven colored squares of pavement. The month of January is garnet red; December is dark brown. "I don't really know where it comes from," she said. "I just know it's always been that way."

—ERICA GOODE, February 1999

A Decade of Discovery Yields a Shock About the Brain

AS SCIENTISTS LOOK BACK at all the discoveries made in the 1990's, the so-called Decade of the Brain, one finding stands out as the most startling and, for many scientists, the most difficult to accept: people are not necessarily born with all the brain cells they will ever have.

In fact, from birth through late adolescence, the brain appears to add billions of new cells, literally constructing its circuits out of freshly made neurons as children and teenagers interact with their environments. In adulthood, the process of adding new cells slows down but does not stop. Mature circuits appear to be maintained by new cell growth well into old age.

Although the Congressionally mandated "Decade" produced many other discoveries, from ways to obtain images of fleeting thoughts inside a person's head to new drugs for a wide variety of mental disorders, the finding that the brain develops and maintains itself by adding new cells is the most revolutionary.

If these findings hold up to further scrutiny, the next decade of brain research promises to generate a total revision of how scientists think human minds are organized and constructed.

Findings have already shed new light on mechanisms of learning, memory and aging in normal brains and suggest daring new ways to treat strokes and other brain disorders. Moreover, they may provide solutions to some abiding mysteries—including the way young children who have half their brains surgically removed to treat severe epilepsy go on to develop normally, as if they had whole brains in half the usual amount of space.

Some researchers have begun isolating special cells that continue to divide and produce new brain tissue, with the hope of implanting such cells into areas of the brain that are damaged by disease or accidents.

For decades, it was axiomatic that people were born with all the brain cells they would ever have. Unlike the bones, the skin, the blood vessels and other body parts, where cells divide throughout life to give rise to new cells, it was believed that the brain did not renew itself.

Though the brain did add vast amounts of new connections early in life and could compensate somewhat for many injuries, it was thought that no one could be expected to grow more brain cells with age. Quite the opposite. People were told that the only thing they could look forward to was gradual mental deterioration as cells died off and were never replenished.

These ideas were so firmly established that many scientists have a hard time believing the findings, reported in the last couple of years by a number of investigators, that the human brain makes new cells after birth, said Dr. Fred H. Gage, a neuroscientist at the Salk Institute in La Jolla, California. Even when they accept the idea that such cells may exist, they argue there is no proof that they do anything important, he said. And those skeptical of the new developments, like Dr. Pasko Rakic, a neuroscientist at Yale, say that if scientists expect others to change longstanding thinking about brain development, the standard of proof must be set very high.

Dr. Per Andersen, a neurobiologist at the University of Oslo in Norway, said neuroscientists had responded to several of the new findings with "resounding silence." This is probably not because of "active neglect," he said, but "it takes some time to let unexpected results sink down in the mutual consciousness of neurobiologists." In short, the new findings are simply too startling and revolutionary to digest all at once.

Dr. Morten Raastad, also from Oslo, compared resistance to the idea of brains' growing new cells to the way scientists once resisted the idea of plate tectonics and continental drift. The theory was first proposed in 1915, but it was not until scientists completed sea-floor magnetism studies in the 1960's that it was accepted, he said.

The traditional view of human brain development is based on experiments done in the mid-1960's on macaque monkeys by Dr. Rakic. He said then that based on available techniques for detecting dividing cells in brain tissue there was no evidence that new cells were being born in the monkey brain. He and others inferred this must be true of all primates, including humans.

According to this theory, brains grow as new connecting fibers, called synapses and dendrites, proliferate around a fixed number of brain cells after birth. Cells not connected into circuits through these growing fibers would die off.

Thus brains develop by pruning and sculpturing, not by building networks with billions of new cells, Dr. Rakic and others theorized.

The fact that many people do not recover the ability to speak or walk after having strokes or other traumatic brain injury cemented the view that adult brains did not add new cells. If they did, people thought, recovery would be more common.

The first crack in this belief occurred in 1965, when scientists reported that new nerve cells were generated in a region of the adult rat brain called the hippocampus. This is where memories for places and things are first formed. A year later, they discovered that new cells were migrating to the olfactory bulb, where smells are decoded.

These researchers identified a zone within two hollow cavities of the rat brain, called ventricles, where new cells are born and then migrate to the brain's interior. The zone contains so-called stem cells that give rise to many other cell types, including neurons and glial cells that nourish neurons.

The new cells seen in the rat brains appear at a higher rate after challenges like intense training, injury or an infection, Dr. Raastad said. Within a few years, researchers found the cells in adult mice, guinea pigs, rabbits and monkeys. In the mid-1980's, other researchers found irrefutable evidence that new cells were born in the brains of adult canaries learning new songs and chickadees that were remembering where they had stashed their winter seeds. But researchers still did not believe that new cells were created in human brains, Dr. Raastad said.

In 1997, Dr. Elizabeth Gould, an assistant professor of neuroscience at Princeton, and colleagues showed that neurogenesis, or the birth of new cells, occurred in the hippocampuses of tree shrews and marmoset monkeys. But Dr. Rakic and others said this was not possible in humans.

In 1998, Dr. Gage demonstrated that the number of brain cells in the hippocampuses of mice raised in stimulating environments increased by 15 percent—and that the cells were born in the ventricle zone.

"This made us go look for the same in humans," Dr. Gage said. Swedish colleagues were using a special substance that integrates into the DNA of dividing cells to track tumor cells in cancer patients, he said. Last year, this substance was found in the hippocampuses of five cancer patients whose brains were dissected immediately after they died.

This was a "thrilling" discovery, Dr. Gage said. It means that the human brain makes new cells in an area already known to be involved in short-term memory. Some sort of neurogenesis may be widespread in the brain and spinal cord for maintenance, he said. Like skin, the brain may be repairing itself all the time. But like a big gash to the skin, a large brain injury like a stroke can overwhelm the repair system.

As for the rest of the brain, including the cortex, where complex functions like language and long-term memories reside, Dr. Gould injected the same dye used in the human experiments into macaque monkey brains. By tracing the chemical, she found that neurons had been born in the ventricles and had migrated into the higher cortex, where they made new axons. They appeared to connect up to local circuitry and perhaps extend into wider circuits, she said, adding that the same might be true for human brains.

But the most surprising finding about new cell growth in the human brain has been virtually ignored by most neuroscientists. This part of the story began more than two decades ago when a young doctor in training, William Rodman Shankle, salvaged a stack of cardboard boxes containing the largest database ever collected on the developing human cerebral cortex. The data had been collected from 1939 to 1967 by Dr. Jesse L. Conel of Boston Children's Hospital, who examined the brains of infants and children up to age 6 who had died from accidents or diseases not affecting brain cells. Before his death, he made more than four million measurements, including the width, thickness and packing density of brain cells at birth and at 1, 3, 6, 15, 24, 48 and 72 months of age.

Dr. Conel published eight volumes of research. Several boxes of his raw data were about to be thrown out—tissue samples and slides already having been discarded—when Dr. Shankle, now a neurologist at the University of California at Irvine, noticed them stacked in a hallway at Boston University and rescued them.

Dr. Conel did not have computer tools to measure exact numbers of cells, Dr. Shankle said, but he did describe, at each age and within 35 brain areas, the appearance of vertical columns of neurons. It is now known that higher brain functions stem from arrays of these columns.

Dr. Shankle and his colleagues re-examined Dr. Conel's data using modern mathematical and computer techniques to allow for cell shrinkage and to distinguish neurons from other kinds of brain cells. They found an astonishingly dynamic pattern in all 35 areas. In each square millimeter of tissue, Dr. Shankle said, the number of neurons rises by a third from birth to 3 months as new cells are added. Then the number plummets back to birth level between 3 and 15 months. After this point, the number increases rapidly, doubling by the age of 6 years. It probably continues increasing, although at a slower rate, up to age 18 or 21, Dr. Shankle said.

The brain enlarges by making new columns, not by making existing ones larger, Dr. Shankle said. "I suspect that a single set of rules constructs all brains," he said. "Children progress through the same stage of development at the same rates independent of their culture."

This rapid growth and construction of brain tissue may help explain why children whose left or right brain hemispheres are removed entirely seem to develop more or less normally, Dr. Shankle said. The rate of growth, or plasticity, is so large early on, they can learn to do most things with their remaining brain tissue.

Dr. Anderson said that Dr. Shankle's findings were "well described and adequately analyzed," and concluded, "I see no major flaws in his handling of the material."

But Dr. William T. Greenough, a University of Illinois neuroscientist, said he was not yet convinced that Dr. Shankle had proved that the growth was from new nerve cells and not from supporting cells called glia. "He may be right," he said, "but the work needs to be replicated."

Meanwhile, Dr. Steven A. Goldman of Cornell Medical Center in New York City is studying human brain tissue removed from epilepsy patients and has found progenitor cells in the ventricles. About 10 percent of cells in this zone are progenitors that give rise to other cell types, he said. This is a trivial number compared with the brain's 100 billion cells, but it may be enough to carry out maintenance and repair of the higher cortex, he said.

The challenge is to make such cells useful, Dr. Goldman said. "We still don't know where they go," he said, "but we do know they're dividing. Some are becoming neurons." If ways could be found to induce their expansion in the human brain, he said, new treatments for a wide variety of brain disorders would be on the near horizon.

—Sandra Blakeslee, January 2000